The Mathematics of Petri Nets

The Mathematics of Petri Nets

CHRISTOPHE REUTENAUER
University of Quebec at Montreal

Translated by
IAIN CRAIG
University of Warwick

036572 31

MATH-STAT.

This edition published 1990 by
Prentice Hall International (UK) Ltd
66 Wood Lane End, Hemel Hempstead
Hertfordshire HP2 4RG
A division of
Simon & Schuster International Group

This book was originally published under
the title *Aspects mathématiques des réseaux de Pétri*
by C. Reutenauer.

Printed and bound in Great Britain by
BPCC Wheatons Ltd, Exeter

Library of Congress Cataloging-in-Publication Data

Reutenauer, Christophe.
   [Aspects mathématiques des réseaux de Petri. English]
   The mathematics of Petri-nets / Christophe Reutenauer:
   translated by Iain Craig.
     p. cm.
   Translation of: Aspects mathématiques des réseaux de Petri.
   Includes bibligraphical references.
   ISBN 0-13-561887-8
   1. Petri nets. 2. Machine theory. I. Title.
QA267.R4713 1990
511.3—dc20
                              89-39986
                                CIP

British Library Cataloguing in Publication Data

Reutenauer, Christopher, *1953-*
   The mathematics of petri-nets.
   1. Systems. Mathematical models: Petri nets
   I. Title
003'.0724

   ISBN 0-13-561887-8

1 2 3 4 5 94 93 92 91 90

ISBN 0-13-561887-8

Contents

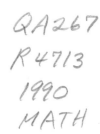

Preface

Petri nets are a tool for modeling communication between parallel processes: they have found considerable application in a number of different areas. Resource allocation problems in information processing systems, communications protocols, production control and process synchronization can be cited as examples of Petri net applications. This diversity of application has encouraged the study of Petri net theory. The amount of activity in Petri net theory has been increasing during the last few years, and there has now developed a considerable body of knowledge on the subject. There are now available several treatments of the basics of Petri net theory and their applications: for example, there are the books by Starke (1981) and Reisig (1982, 1985).

A book which presents a detailed exposition of advanced results in Petri net theory has been lacking, however. In particular, there is no book that deals with decidability issues, the most important of which is the celebrated accessibility theorem. The aim of the present monograph is to fill this gap, at least in part. The principal result presented here is the proof of the accessibility problem for Petri nets: this book also contains closely related decidability results.

The accessibility problem can be stated very simply: it consists of determining whether a given network marking can be derived from an initial marking. The importance of this question is clear, both from the theoretical and from the practical standpoints: it is for this reason that the question of knowing whether there exists an algorithm to answer this question was asked early on in Petri net research. The first mention of the problem dates back as far as 1969 (Karp and Miller). Various researchers have tackled the problem and have presented partial solutions – for example, van Leeuwen (1974) and Hopcroft and Pansiot (1979). One of the first decidability proofs (that of Sacerdoti and Tenney, 1977) did not stand up to closer examination. Two complete proofs were published independently by Mayr (1981, 1984) and by Kosaraju (1982).

The presentation of the latter proof is the subject matter of this book. The proof is extremely complex and requires concepts and results from different areas of mathematics, such as arithmetic, logic and formal language theory. Because of the desire to give a fairly elementary account, we have included material necessary for a totally independent presentation (it is for this reason that all the concepts employed in this book are fully defined and explored). This aim accounts, for example, for the treatment of semi-linear sets and of natural-valued linear equations (this last theory is equivalent to Presburger arithmetic – cf. Ginsburg and Spanier, 1966). All of the theorems necessary for the development of the central theme of this book are stated and proven. We develop several theorems in Chapters Two and Three, but we will concentrate only on those that are indispensable to the development of the central argument. However, as this book is written under the general umbrella of algorithmic decidability, all proofs are constructive, and, although this fact is never explicitly stated, the reader will easily be convinced of the fact.

The first chapter serves to introduce Petri nets, vector addition systems and vector addition systems with states (VASS). We show the equivalence of these concepts.

Chapter Two presents some results about graphs and ordered sets. It also presents the necessary and sufficient conditions for the correspondence between a bag and a path. König's lemma and the finiteness conditions for certain ordered sets (in particular, the condition which assures the termination of Kosaraju's algorithm) are presented.

Chapter Three is given over to the study of rational subsets in free and free commutative monoids, and to the rationality of sets of paths in a graph (this is part of Kleene's theorem). Semi-linear sets, closure by intersection, natural solutions of linear equations and minima in ideals over N^k are also examined.

Chapter Four begins with the re-introduction of vector addition systems with states, together with all the necessary terminology. The central part of the chapter relates to Karp and Miller's VASS tree, and the tricky theorem (due to Kosaraju) which gives the sufficient condition for the existence of a positive walk through a VASS – in a similar way, this is a sufficient condition for the accessibility of one marking from another in a Petri net.

In the following chapter, we present the idea of a VASS chain, with a view to proving the decidability of accessibility which is the main goal of the book. All of the concepts previously introduced are used to this end.

Chapter Six is supplementary. We give the necessary condition (due to Valk and Vidal-Naquet, 1981) for the rationality of a Petri net language. Then, we prove the undecidability of the equivalence problem for accessibility sets, as well as the liveness problem (results due to Hack).

The majority of this book was the subject of a DEA course at the Institut de Programmation de l'Université Paris VI in 1984-5, which was given jointly with Jacques Sakarovitch, whom I would like to thank for our discussions and for his course notes. Equally, I would like to thank Jean Berstel, who lent me the notes for his course at the University of Quebec at Montreal, as well as Fabienne Romian and Guy Vidal-Naquet. The production of this book was made possible by help from PRC Mathématique de l'Informatique. The book was written at the Laboratoire d'Informatique Théoretique et de Programmation, using LaTeX. The proofs were read by Rosa de Marchi, and the manuscript benefited from comments by J. Sakarovitch, J. Berstel, Y. Legrandgérard, C. Simian, J. Dupont, S. Perras, and all those whom I met in the machine room: I am grateful to them all.

Christophe Reutenauer
Paris and Montreal, 1985-87

CHAPTER ONE

Petri Nets
and Vector Addition Systems

In this short chapter we introduce the ideas behind Petri nets, vector addition systems and vector addition systems with states (VASS). We will show how these three models are equivalent and that each can simulate the others. These constructions will serve us at various points in this book, notably when moving from a VASS result to an analogous Petri net result.

1.1 Petri Nets

A Petri net is a 4-tuple $(P, T, Pre, Post)$ where the following conditions hold:
- P is a finite set whose elements are called *places.*
- T is a finite set whose elements are called *transitions.*
- *Pre* is a function $P \times T \to \mathbf{N}$ called the *forward incidence function.*
- *Post* is a function $T \times P \to \mathbf{N}$ called the *backward incidence function.*

A *marking* of a Petri net R is a mapping $M: P \to \mathbf{N}$. We will say that $M(p)$ is the number of markers at place p for the marking M. We will often identify a marking with the vector it defines in $\mathbf{N}^P$. A *marked net* is a pair (R, M_0) composed of a Petri net and a marking: the latter will be called the *initial marking.*

We represent a Petri net graphically by an oriented bipartite graph, whose vertices are the places and transitions of the net. For each place p, and each transition t, there are $Pre(p, t)$ arcs from p to t, and $Post(t, p)$ arcs from t to p. In general, a place is represented by a circle and a transition by a rectangle.

1

Furthermore, rather than show n arcs from p to t (or from t to p) we will represent them by a single arc labeled n. A marking M is represented in the graph by $M(p)$ dots at each place p.

Example 1

The places in figure 1.1 are $P = \{1, 2, 3, 4, 5\}$, and the transitions are $T = \{a, b, c, d\}$. We have:

$Pre(1, a) = 1$

$Pre(1, c) = 1 \qquad Post(b, 1) = 1$

$Pre(2, b) = 1 \qquad Post(b, 5) = 1$

$Pre(2, d) = 1 \qquad Post(c, 1) = 1$

$Pre(3, d) = 1 \qquad Post(c, 3) = 1$

$Pre(4, c) = 1 \qquad Post(d, 2) = 1$

$Pre(2, c) = 1 \qquad Post(d, 4) = 2$

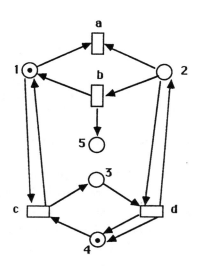

Figure 1.1

The other $Pre(i, x)$ and $Post(x, i)$ are 0. The marking M represented here is $M(1) = 1 = M(4)$, $M(2) = 0 = M(3) = M(5)$.

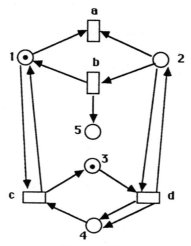

Figure 1.2

Given a net R and a marking M, we say that a transition t is *M-enabled* or that t is *fireable at M* if:

$\forall p \in P, M(p) \geq Pre(p,t)$

which we denote by:

$M(t >$

or by:

$M.t \neq 0$

If t is M-enabled, let M' be the marking defined by:

$M'(p) = M(p) - Pre(p,k) + Post(p,t)$

We say that t *fires* from M to M', and that M' is obtained from M after the *firing* of t. We write this:

$M(t > M'$ or $M.t = M'$

In the example above, transition c is *M-enabled*, because for each place p, the number of arrows $p \to c$ is less than or equal to the number of markers $M(p)$ at place p. After p fires, the net is as shown in figure 1.2.

The other transitions are not fireable for M, since $M(2) = 0$ and $Pre(2, a) = Pre(2, b) = Pre(2, d) = 1 > M(2)$.

Given a Petri net R, a sequence $s = (t_1, \ldots, t_m)$ of transitions, and a marking M, we say that the sequence is *fireable for M* or that it is *M-enabled* if there exists a sequence of markings $M_0, \ldots, M_n$, with $M_0 = M$, such that for each index i between 1 and n we have:

$$M_{i-1}(t_i > M_i$$

In other words, if t_1 is fireable for M, and, after firing, gives M, t_2 is fireable for M_1 and gives M_2 after firing, and so on. As above, this is denoted:

$$M(s > M_n$$

or as:

$$M \cdot s = M_n$$

We say that M_n is *reachable from M*. The reachability set of a marked net (R, M_0) is the set of markings accessible from M_0.

The *reachability problem* for a Petri net is as follows. Given a net R and two markings M and M', does there exist a sequence s of transitions, such that s is fireable, and such that $M(s \geq M'$?

We will give the solution to this problem in the course of this book, and will show that it is decidable.

Given a Petri net R and an initial marking M, a transition t is *live* for the marked net (R, M_0) if, for every marking M accessible from M_0, there exists a transition sequence s containing the transition t at least once, such that s is fireable at M.

A marked net (R, M_0) is called *live* if all of its transitions are live. We will see that the problem of deciding whether a transition is live reduces to the accessibility problem, and that it is also decidable.

1.2 Vector Addition Systems

A *vector addition system* is given by an integer m and a finite set V of vectors in $\mathbf{Z}^m$.

Given such a system, and two points x and y, in $\mathbf{N}^m$, we say that y is *accessible from x* for the system under consideration if there exists a sequence of vectors $v_1, \ldots, v_n$ in V (the sequence may, possibly, contain repetitions), such that

$$y = x + v_1 + \cdots + v_n \tag{1}$$

and such that for $i = 1$ to n we have:

$$x + v_1 + \cdots + v_i \in \mathbf{N}^m \tag{2}$$

In other words, this states that y can be reached from x by adding the vectors in V. The vectors may be repeated if necessary, but the entire operation always remains in $\mathbf{N}^m$.

Example 2

We take $m = 2$, $V = \{v_1, v_2, v_3\}$ with $v_1 = (1, -2)$, $v_2 = (-1, 1)$, $v_3 = (2, 1)$, as shown in figure 1.3.

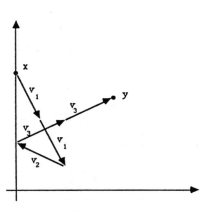

Figure 1.3

Here, $y = (4, 4)$ is accessible from $x = (0, 5)$.

The reachability problem for vector addition systems can easily be mapped onto the analogous problem for Petri nets. To do this we simulate the vector addition system $V = \{v_1, \ldots, v_k\}$ in the following way. For the net we take $R = (P, T, Pre, Post)$, where

$$P = \{1, \ldots, m\}$$
$$T = \{t_1, \ldots, t_k\}$$
$$Pre(j, t_i) = -(v_i)_j, \text{ if } (v_i)_j < 0, = 0 \text{ otherwise}$$
$$Post(t_i, j) = (v_i)_j, \text{ if } (v_i)_j > 0, = 0 \text{ otherwise.}$$

It is easy to see that y is accessible from x for the system V if and only if the marking M' is accessible from M for the net R, where M and M' are defined by:

$$M(j) = x_j, M'(j) = y_j \ (j = 1, \ldots, m)$$

∎

Example 3

The system of example 2 is simulated by the net shown in figure 1.4. The marking shown corresponds to $x = (0, 5)$.

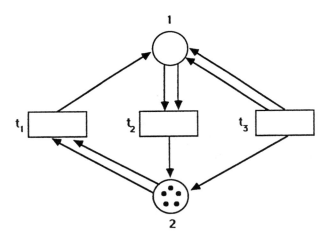

Figure 1.4

∎

Conversely, we can simulate each Petri net by a vector addition system. We proceed indirectly by generalizing these systems slightly. Intuitively, a *vector addition system with states* (VASS) is a vector addition system in which each addition is controlled by a finite-state automaton: thus, each vector v in the system can only be applied when the system is in a certain state (which depends upon v), and the addition of v moves the system into another state (which also depends upon v).

Formally, a VASS is given by a finite, oriented graph $G = (Q, A)$, by an integer $m \geq 1$ and a mapping $v : A \to \mathbf{Z}^m$. Here Q is the set of *vertices* or *states* of G, A is a set of *arcs* or *edges*, and $v(a)$ is the *valuation* or *labeling* of the arc a.

A *configuration* of G is a pair (q, x) composed of a state in G and a point in $\mathbf{N}^m$. A configuration (q', y) is *(positively) accessible* from (q, x) if there exists a sequence $a_1, \ldots, a_n$ of consecutive arcs such that q is the start of a_1 and q' is the end of a_n, and such that:

$$y = x + v(a_1) + \cdots + v(a_n)$$

and such that for all $i = 1, \ldots, n$, we have:

$$x + v(a_1) + \cdots + v(a_n) \in \mathbf{N}^m$$

It is clear that VASS concepts generalize to those of vector addition systems: a vector addition system is merely a VASS with only one state p, where each vector v in V defines a loop $p \to p$ labeled by v.

The Petri net $R = (P, T, Pre, Post)$, $P = \{1, \ldots, m\}$ can be simulated by the VASS $G = (Q, A)$ defined by the following conditions:

- $Q = \{q\} \cup T$.
- $A = T_1 \cup T_2$, where T_1 and T_2 are disjoint copies of T. For each arc t in T, the arc t_1 goes from q to t, and t_2 goes from t to q.
- The valuation v is defined for each t in T by $v(t_1)_j = -Pre(i, t)$ and $v(t_2)_i = Post(t, i)$.

Example 4

The VASS associated by the above construction with the netw in example 1 is shown in figure 1.5: valuations are shown on each arc.

■

It is clear that, with this construction, the reachability problem for Petri nets reduces to the analogous one for VASSs. In effect, given two markings M and M' for a net R, the marking M' is accessible from M if and only if the configuration (q, y) is accessible from (q, x) in the VASS

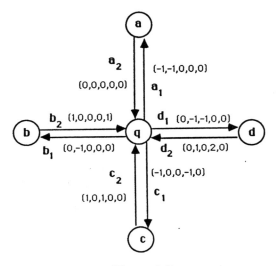

Figure 1.5

given above, where x and y are defined by $\forall j = 1, \ldots, m, x_j = M(j), y_j = M'(j)$.

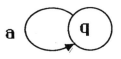

Figure 1.6

We now show how one can simulate a VASS with a vector addition system. We first take the case of a VASS without loops (as in the VASS of example 4 – see figure 1.6). For this, a loop is simulated as in figure 1.7, where q' is a new state and a' is a new arc with $v(a') = (0, \ldots, 0)$.

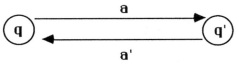

Figure 1.7

Having done this, we then have a VASS $G = (Q, A)$ which is labeled in $\mathbf{Z}^m$, which has no loops. We now go on to construct a vector addition system V in $\mathbf{Z}^m \times \mathbf{Z}^Q$ in the following way: for each arc $p \xrightarrow{a} q$ in G, we define the following vector in V:

$$(v(a), u) \in \mathbf{Z}^m \times \mathbf{Z}^Q$$

where u is given by $u_j = -1$ if $j = p$, $u_j = 1$ if $j = q$, $u_j = 0$ otherwise.

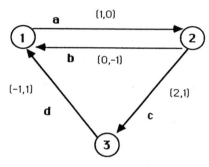

Figure 1.8

It is then easily verified that the configuration (q, y) of the VASS is accessible from (p, x) if and only if, in V, the point (y, w) is accessible from (x, u), where u is defined as follows:

$$u_j = \begin{cases} 1, & \text{if } j = p \\ 0, & \text{otherwise} \end{cases}$$

$$w_j = \begin{cases} 1, & \text{if } j = q \\ 0, & \text{otherwise} \end{cases}$$

Example 5

The VASS shown in figure 1.8 is simulated by the vector addition system V in $\mathbf{Z}^2 \times \mathbf{Z}^3$ with $v = \{a, b, c, d\}$, and $a = (1, 0, -1, 1, 0)$, $b = (0, -1, 1, -1, 0)$, $c = (2, 1, 0, -1, -1)$, and $d = (-1, 1, 1, 0, -1)$.

∎

Remark

A construction by Hopcroft and Pansiot (1979) shows that each VASS labeled by elements of $\mathbf{Z}^m$ can be simulated by a vector addition system in $\mathbf{Z}^m + 3$ (cf. exercise 1.5)

Exercises

1.1 Let R be a Petri net, and let M, $M' \in \mathbf{N}^p$ be two markings. Show that if M' is accessible from M, then the vector $M' - M$ belongs to the subgroup of $\mathbf{Z}^P$ generated by the vectors $V_t\,(t \in T)$, where $V_i = (Post(p,t) - Pre(p,t))$, $p \in P$.

1.2 Let A_t and B_t be vectors in $\mathbf{N}^k$, where t ranges over a finite set T. We denote by $\rightarrow$ the least binary relation in $\mathbf{N}^k$ which is reflexive, transitive, compatible with addition over $\mathbf{N}^k$, and which contains the relations $A_i \rightarrow B_i$. Construct a Petri net R with $\{1, \ldots, k\}$ as its set of places, such that, for all markings $M, M' \in \mathbf{N}^k$, M' is accessible from M if and only if $M \rightarrow M'$. (Compatible with addition means that $A \rightarrow B$ implies $A + U \rightarrow B + U$, for all vectors A, B, U in $\mathbf{N}^k$.)

1.3 Give a criterion analogous to that of exercise 1.1 for the reachability problem for vector addition systems.

1.4 Show that a Petri net is obtained from a vector addition system by the construction of section 1.2 if and only if it does not contain a subgraph of the form shown in figure 1.9.

Figure 1.9

1.5 Let $G = (Q, A)$ be a VASS labeled by $v: A \rightarrow \mathbf{Z}^m$, with $Q = \{q_1, \ldots, q_k\}$. We define a vector addition system in $\mathbf{Z}^m + 3$ by

$a_i = i$, $b_i = (k+1)(k+1-i)$, $1 \leq i \leq k$, and:

$V =$

$$\{(0,\ldots,0,-a_i, a_{k-i+1} - b_i, b_{k-i+1}) | 1 \leq i \leq k\}$$
$$\cup \{(0,\ldots,0), b_i, -a_{k-i+1}, a_i - b_{k-i+1} | 1 \leq i \leq k\}$$
$$\cup \{(v(a), a_j, -b_i, b_j, -a_i) | a : q_i \to q_j \in A\}$$

Show that this system simulates G. Hint: the configuration (q_i, x) will be simulated by the vector $(x, a_i, b_i, 0)$ (see Hopcroft and Pansiot, 1979, lemma 2.1).

CHAPTER TWO

Graphs and Ordered Sets

This chapter is given over to the proof of some results about graphs and about ordered sets: these results will be of use later. We will see, first of all, a necessary and sufficient condition that a multi-set of edges in a graph corresponds to a path in the graph – this is an extension of the existence condition for Euler trails. In the short second section, we prove König's lemma: in every infinite tree which is locally finite, there is an infinite path. In the last section, we show that the lexicographic ordering on $\mathbf{N}^k$ is Artin (i.e., there is no infinite sequence that is strictly decreasing). Next, we extend the Artin ordering on a set A to the free monoid generated by A (in a way reminiscent of the ordering on subwords), and we show that the new ordering remains Artin. This shows the termination of the Kosaraju algorithm.

2.1 Commutative Image of a Path in a Graph

Here, we consider a finite, *oriented graph* $G = (S, A)$ where S is the set of vertices and A the set of *arcs* or *edges*. We denote by $\alpha(a)$ and $\omega(a)$ the *start* and *end* of an arc $a \in A$. Recall that a path c in G is a finite sequence of arcs $c = (a_1, \ldots, a_n)$ or $c = a_1 \cdots a_n$ satisfying the condition that $\omega(a_i) = \alpha(a_{i+1})$ for $i = 1, \ldots, n - 1$. We will call $\alpha(c)$ and $\omega(c)$ the *start* and *end* of the path c, i.e., $\alpha(c) = \alpha(a_i)$ and $\omega(c) = \omega(a_n)$. We will also write $c \colon \alpha(c) \to \omega(c)$ or $\alpha(c) \xrightarrow{c} \omega(c)$ to emphasize this. A path c is *closed* if $\alpha(c) = \omega(c)$. The *length* of a path c, denoted by $|c|$ is n. We also consider *empty* paths: they are always closed and have a length of 0. There is an empty path for each vertex in the graph. Two paths c

and c' are said to be *consecutive* if $\omega(c) = \alpha(c')$. We will denote by cc' the path obtained by *concatenating* the paths c and c'.

A graph is said to be *strongly connected* if, for each pair of vertices s and t, there exists a path $s \to t$. In the same way, a graph is said to be *(weakly) connected* if the unoriented graph underlying G (i.e., the graph obtained from G by ignoring orientations, or by adding an arc $t \to s$ for each arc $s \to t$ in G) is strongly connected. An *isolated point* in G is a vertex s with no incident arcs (i.e., there is no arc whose start or end is s).

We define, more generally, the *strongly connected components* of a graph G. The relation over the set of vertices: 'there exists a path $s \to t$ and a path $t \to s$' is clearly an equivalence relation. A strongly connected component of G is an equivalence class for this relation. Observe that a graph is strongly connected if and only if it has exactly one strongly connected component.

Below, we will consider the set $\mathbf{Z}^H$ of functions from a finite set H to $\mathbf{Z}$, and similarly for the set $\mathbf{N}^H$. The function whose value is 1 at h_0 and 0 at all $h \neq h_0$ will be written simply as h_0. Thus, each element v of $\mathbf{Z}^H$ (respectively, $\mathbf{N}^H$) will be written in a unique fashion as a linear combination with coefficients v_h in $\mathbf{Z}$ (respectively, in $\mathbf{N}$):

$$v = \sum_{h \in H} v_h h$$

which therefore denotes the function v which associates v_h with each h. In other words, $\mathbf{Z}^H$ is considered here as the free $\mathbf{Z}$-module with base H. It is also convenient to see $\mathbf{Z}^H$ as the set of sequences of elements of $\mathbf{Z}$ indexed by H, i.e., as vectors of the form:

$$v = (v_h), \quad h \in H$$

In this way, H represents the canonical base of $\mathbf{Z}^H$, i.e., each h_0 in H is thought of as the vector

$$(\delta_{h,h_0}), \quad h \in H$$

where δ is the Kronecker delta.

The *result* of a path c in a graph G determines the path's start and end. By definition, the result of c is:

$$\beta(c) = \omega(c) - \alpha(c) \in \mathbf{Z}^S$$

using the conventions given above. In particular, $\beta(c)$ *is zero if and only if c is a closed path.*

The *commutative image* of a path c will be used to determine the number of times that each arc in the graph occurs in c. To do this we denote by $|c|_a$ the number of occurrences of the arc a in the path c. The commutative image of c is then:

$$\gamma(c) = \sum_{a \in A} |c|_a \quad a \in \mathbf{Z}^A$$

using the above conventions.

Example 1

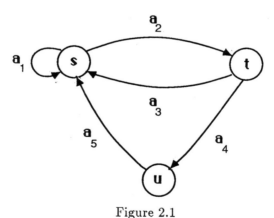

Figure 2.1

Let G be the graph shown in figure 2.1. Let c be the path $a_1 a_1 a_1 a_2 a_3 a_2$ $a_3\, a_2$. Then the value of c is $\beta(c) = t - s$, and its commutative image is

$$\gamma(c) = 3a_1 + 3a_2 + 2a_3$$

Using our conventions, $\gamma(c)$ will also be represented by the vector

$$\gamma(c) = (3, 3, 2, 0, 0)$$

in $\mathbf{N}^A = \mathbf{N}^5$.

If c and c' are two consecutive paths, it is clear that:

$$\beta(cc') = \beta(c) + \beta(c') \text{ and } \gamma(cc') = \gamma(x) + \gamma(c')$$

In particular, if $c = a_1 \cdots a_n \ (a_i \in A)$, then:

$$\beta(c) = \sum_{i=1}^{n} \beta(a_i) = \sum_{a \in A} |c|_a \beta(a) \tag{1}$$

This shows that *the result of a path only depends on its commutative image.*

■

We now show under which conditions an element of $\mathbf{N}^A$ is the commutative image of a path. First, we define the result, which we also denote by β, of an element $v = \sum_{a \in A} v_a \, a \in \mathbf{N}^A$ as:

$$\beta(v) = \sum_{a \in A} v_a (\omega(a) - \alpha(a))$$

$$= \sum_{a \in A} v_a \beta(a)$$

For every path c, we then have $\beta(c) = \beta(\gamma(c))$ as in (1). Moreover, for v in $\mathbf{N}^A$ as before, we define the restriction of the graph G to v as

$$G|v = (S', A')$$

where $A' = \{a \in A | v_a > 0\}$ and $S' = \{s \in S | \exists a \in A', \alpha(a) = s \text{ or } \omega(a) = s\}$. In other words, $G|v$ is the graph obtained by removing from G all those arcs which do not occur in v.

For example 1, let $v = 3a_1 + 3a_2 + 2a_3$. We have $\beta(v) = t - s$. Furthermore $G|s$ is the graph shown in figure 2.2. As we have seen, v is the commutative image of the path $a_1 a_1 a_1 a_2 a_3 a_2 a_3 a_2$. But v is also the commutative image of the path $a_1 a_2 a_3 a_1 a_2 a_3 a_1 a_2$.

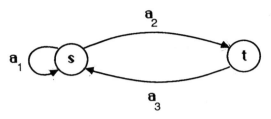

Figure 2.2

If v is the commutative image of a path, it is clear that $G|v$ is a connected graph. Furthermore, the value of v has the form $t-s$, for two (eventually identical) vertices s and t in the graph. We will see that these two conditions characterize the commutative images of paths.

Theorem 2.1.   *Let $G = (S, A)$ be an oriented, finite graph and let v be an element of $\mathbf{N}^A$. Then v is the commutative image of a path if and only if the two following conditions are met:*
   (i) The graph $G|v$ is connected.
   (ii) The value of v is the difference of two vertices in the graph.
If, furthermore, the value of v is zero, then for each vertex s in $G|v$ there exists a path $s \rightarrow s$ with commutative image v.

The proof is constructive. We begin by finding a set of paths for which v is the commutative image; we connect two of these paths to form a single one, and then iterate the process until only one path remains.

PROOF. 1. It is enough to show that conditions (i) and (ii) are sufficient. So, let v be a path satisfying (i) and (ii). We can assume that $v \neq 0$. Then there must exist a set $\{c_1, \ldots, c_n\}$ of non-empty paths such that $\sum_{i=1}^{n} \gamma(c_i) = v$ (take the sequence formed from v_a times arc a, v_b times b, and so on, for all arcs a, b, ... in A). We select such a set minimal n and show that $n = 1$, hence $v = \gamma(c_1)$.

Suppose that $c_1, \ldots, c_p$ are not closed and that $c_{p+1}, \ldots, c_n$ are closed. If $p \geq 2$, since $\beta(c_{p+1}) = \ldots = \beta(c_n) = 0$, we have:

$$t - s = \beta(v) = \sum_{i=1}^{p} \beta(c_i)$$

$$= \sum_{i=1}^{p} (\omega(c_i) - \alpha(c_i))$$

Since the second sum has $p \geq 2$ terms, there exists $j \neq k$, $1 \leq j$, $k \leq p$ such that $\omega(c_j) = \alpha(c_k)$. We can then replace $\{c_1, \ldots, c_n\}$ by the smaller set obtained by replacing c_j, c_k by $c_j c_k$. This contradicts the minimality of n, and it follows that $p = 0$ or $p = 1$.

The set $\{c_1, \ldots, c_n\}$ is represented schematically in figure 2.3 (only c_1 is perhaps not a closed path). But, if $n \geq 2$, because $G|v$ is connected, two of the paths shown above have a common vertex, and we obtain a smaller set by virtue of the following fact.

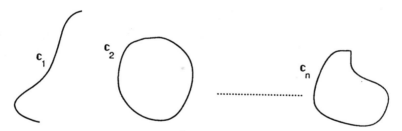

Figure 2.3

FACT: If x and y are two paths having a common vertex, and if y is closed, then there exists a path c such that $\gamma(c) = \gamma(x) + \gamma(y)$.

Indeed, let s be the common vertex. Then $x = x_1 x_2$ and $y = y_1 y_2$, with $s = \omega(x_1) = \alpha(x_2) = \omega(y_1) = \alpha(y_2)$. As y is closed, $y_2 y_1$ is a path $s \rightarrow s$, so $c = x_1 y_2 y_1 x_2$ is a path satisfying $\gamma(c) = \gamma(x) + \gamma(y)$.

It follows by minimality of n, that $n = 1$, and that v is the commutative image of c_1.

2. Suppose, now that $\beta(v) = 0$, and that $G|v$ is connected. Since $v = \gamma(c_1)$ and $\beta(c_1) = \beta(v) = 0$, the path c_1 is necessarily closed. Let s be

a vertex in the graph $G|v$. Then there exists an edge a such that $v_a > 0$, hence $|c_1|_a > 0$, and such that $s = \alpha(a)$ or $s = \omega(a)$. Since c_1 is closed, we have $c_1 = xy$ with $\omega(x) = \alpha(y) = s$. So, yx is a closed path $s \to s$ with commutative image v. □

Theorem 2.1 is useful most of all because of its corollary.

Corollary 2.2.   *Let $G = (S, A)$ be a finite, oriented graph such that its restriction to non-isolated vertices is connected. Let $v \in \mathbf{N}^A$ have result 0 and such that $v \geq (1, \ldots, 1)$ – i.e., $v_a \geq 1$ for all $a \in A$. There therefore exists a path $s \to s$ with commutative image v for each non-isolated vertex of G.*

The reader will easily prove that, given the above assumptions, the graph $G \backslash \{isolated\ points\}$ is strongly connected (cf. exercise 2.1).

PROOF. Since $v \geq (1, \ldots, 1)$, each arc in G occurs in v, so $G|v = G \backslash \{isolated\ points\}$. Since $G|v$ is connected and $\beta(v) = 0$, theorem 1.1 applies, and there exists a path $s \to s$ with commutative image v for each vertex s in G. □

Theorem 2.1 is, in fact, a generalization of the classical condition for the existence of Euler trails in an oriented graph. Recall that an Euler trail in a graph is a path which contains each arc of the graph exactly once.

Corollary 2.3.   *For a connected, finite, oriented graph to contain a closed Euler trail, it is necessary and sufficient that, for each vertex in the graph, there are as many arcs entering as leaving.*

The proof is left to the reader (cf. exercises 2.2 and 2.3).

2.2 König's Lemma

Recall that a *tree* is an oriented graph $T = (S, A, r)$, where S is the set of vertices, A the set of arcs, and r is a distinguished vertex (called the *root*), such that: (1) no arc ends at r; (2) there is exactly one arc ending at each vertex $\neq r$; (3) each vertex is accessible from the root.

Let $a: s \to t$ be an arc. Call s the *father* of t, and t the *son* of s. If $c: s \to t$ is a path, then s is an *ancestor* of t and t a *descendant* or *successor* of s.

A tree is said to be *locally finite* if each vertex has only a finite number of sons, or, in an equivalent fashion, if there is no more than a finite number of arcs starting at any given vertex.

An infinite path through a graph is a sequence of arcs (a_n), $n \in \mathbf{N}$ such that $\forall n \in \mathbf{N}, \omega(a_n) = \alpha(a_{n+1})$.

The following lemma, due to König (1950, Satz 6.6, p. 81) will be useful at two points in this book.

Theorem 2.4. *Let $T = (S, A, r)$ be an infinite tree (i.e., A is infinite) which is locally finite. Then there exists an infinite path in T.*

It might be useful if we note that the mapping $a \mapsto \omega(a)$ is a bijection from A to $S \backslash r$; so A and S are both infinite.

PROOF. *(See also exercise 2.5.)* If s is a vertex of T, we use $T(s)$ to denote the subtree of T rooted at s: i.e., the subgraph of T whose vertices are s and its successors. If $T(s)$ is infinite, and $t_1, \ldots, t_n$ are the sons of s, it is clear that one of its subtrees $T(t_i)$ is infinite. Moreover, $T = T(r)$ is infinite. This shows that there exists an infinite sequence of vertices (s_n) with $s_0 = r$, such that s_{n+1} is the son of s_n, and such that $T(s_n)$ is an infinite tree. The sequence (s_n) then defines an infinite path in T. □

2.3 Ordered Sets

Let $\mathcal{N} = \mathbf{N} \cup \{\infty\}$. The usual ordering on $\mathbf{N}$ is extended to $\mathcal{N}$ using the condition that $n < \infty$ for each natural number n. The *natural ordering* on $\mathcal{N}^k$ is $\leq$ (a partial ordering if $k \geq 2$), defined coordinatewise, that is;

$$u \leq v \Leftrightarrow \forall i \in \{1, \ldots, k\}, u_i \leq v_k$$

where u_i is the ith coordinate of u. This ordering possesses the properties stated in the following theorem.

Theorem 2.5. *(i) Each sequence in $\mathcal{N}^k$ has an increasing sequence; (ii) Every set formed from pairwise incomparable elements of $\mathcal{N}^k$ is finite.*

Recall that a sequence u_n is said to be *increasing* (respectively, *strictly increasing*), if for each integer n, $u_n \le u_{n+1}$ (respectively, $u_n < u_{n+1}$).

PROOF. (i) Let (u_n), $n \in \mathbf{N}$ be a sequence of values in $\mathcal{N}^k$. We begin with the case $k = 1$. If there is an infinite number of n such that $u_n = \infty$, we can immediately form a constant sequence (u_n). If not, we replace (u_n) by the subsequence (v_n) obtained by replacing the n by $u_n = \infty$. Then, either (v_n) is bounded and we can extract a constant sequence, or else it is not and we can extract a strictly increasing sequence.

For $k \ge 2$, we begin by extracting a subsequence which is increasing in its first coordinate. We then argue by induction on k, using the fact that a subsequence of an increasing sequence is always increasing.

(ii) If such a set were infinite, we could create a sequence contradicting (i). □

Let $(A, \le)$ be an ordered set. The ordering is called *Artin* if there exists no strictly decreasing sequence (a_n) in A. From an ordered Artin set A, we will construct another ordered set and will show that it, too, is Artin. This is the crucial point in showing the termination of Kosaraju's algorithm.

Let A^* be the *free monoid* generated by A, i.e., the set of *words* over A including the empty word (which will be denoted by 1). We write:

$$u \rightarrow v$$

to indicate that u and v are two words of A^* such that $\exists x, y, w \in A^*, w \ne 1$, $\exists a \in A$ such that $u = xay$, $v = xwy$, and for each *letter* (= element of A) b which occurs in w, we have $a > b$. We use $\overset{*}{\rightarrow}$ to denote the reflexive, transitive closure of $\rightarrow$.

Theorem 2.6. *If A is an Artin ordered set, then the relation defined by $u \ge v$, if $u \overset{*}{\rightarrow} v$, is an Artin ordering over A^*.*

The proof makes use of König's lemma, and is illustrated by the following example.

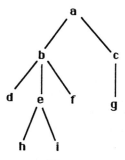

Figure 2.4

Example 2

With $A = \{a, b, \ldots\}$, and an ordering $>$ such that $a > b, c$; $b > d, e, f$; $e > h, i$; $c > g$, we consider the sequence $u_0 \to u_1 \to u_2 \to u_3 \to u_3 \to u_4$, where $u_0 = \underline{a}$, $u_1 = \underline{b}c$, $u_2 = d\underline{e}fc$, $u_3 = dhif\underline{c}$, and $u_4 = dhifg$, where replacement of the underlined letter gives the following word according to the definition of $\to$. The tree in figure 2.4 is associated with this sequence.

■

By the definition of $\to$, each vertex in the tree is larger than each of its sons. If there exists an infinite sequence (u_n) such that $u_n \to u_{n+1}$ for all n, an infinite tree can be formed, and this tree has an infinite path – we could then infer an infinite, strictly decreasing sequence in A.

PROOF OF THEOREM 2.6. It suffices to show that there is no infinite sequence (u_n) of words in A^* such that $\forall n \in \mathbf{N}, u_n \to u_{n+1}$. By way of contradiction, suppose that there exists such a sequence. For technical reasons, it is convenient to suppose that u_0 is a letter, if necessary by adding one to A that is larger than any other, and adding it to the sequence as the first term.

We know that, for all n, there exist words x_n, y_n and w_n, and that there exists a letter a_n such that $w_n \neq 1$, $u_n = x_n a_n y_n$, $u_{n+1} = x_n w_n y_n$, and such that a_n is larger than any letter in w_n. Let E be the set of *letter occurrences* in the words u_n. Formally, this is the set:

$$E = \{(n, i) | n \geq 0, 1 \leq i \leq |u_n|\}$$

where $|u|$ is the length of word u. The set S of word occurrences w_n, that is:

$$S = \{(n,i)|n \geq 0, 1 \leq i \leq |w_n|\}$$

can be thought of as a subset of E, since w_n is a factor of u_{n+1}. We will tacitly use this injection of S into E.

The *origin* of o in E is an element of E, and is defined by: if $o = (0,1)$, $origin(o) = o$; if $o = (n,i)$, $n \geq 1$, is an occurrence in $u_n = x_{n-1}w_{n-1}y_{n-1}$, then it is in w_{n-1}, in x_{n-1} or in y_{n-1}. In the first case, $origin(o) = o$; in the second and third case, since $u_{n-1} = x_{n-1}w_{n-1}y_{n-1}$, o defines an occurrence o' for u_{n-1} in either x_{n-1} or y_{n-1}; then, $origin(o) = origin(o')$.

We now define a tree T with root $(0,1)$ whose other vertices are elements of S: this tree will, therefore, be infinite. For s in S, s is an occurrence in some w_n; its father in T is, by definition, the origin of the occurrence a_n in $u_n = x_n a_n y_n$.

We observe that if (n,i) is the father of (p,j), and, if a and b are the letters that are respectively underlying these letter occurrences, then $a > b$. Now, by König's lemma, there exists an infinite branch in T (theorem 2.4). We obtain, therefore, an infinite decreasing sequence in A: a contradiction. □

We will need to apply this theorem to the case of the set $A = \mathbf{N}^3$ which is ordered *lexicographically*. Recall that the *lexicographic ordering* over $\mathbf{N}^k$ is defined by $x < y$ if and only if there exists an i, $1 \leq i \leq k$, such that for all j satisfying $1 \leq j \leq i-1$, $x_j = y_j$, and $x_i < y_i$.

Proposition 2.7.   *The lexicographical ordering on $\mathcal{N}^k$ is Artin.*

Note that when $\mathbf{N}^k$ is ordered lexicographically, the set $\{x|x \leq a\}$ is not always finite (cf. exercise 2.6).

PROOF. Let $x_1, \ldots, x_n, \ldots$ be an infinite, strictly decreasing sequence in $\mathbf{N}^k$. We can extract an infinite sequence which is constant in its first coordinate, but this leads immediately to the case for $k-1$. □

Exercises

2.1 Let G be an oriented graph containing a path that contains each edge twice. Show that the graph $G\backslash\{isolated\ points\}$ is strongly connected.

2.2 Show that an oriented graph contains an Euler trail if and only if the sum $\sum_{a\in A}(\omega(a) - \alpha(a))$ is equal to the difference between two vertices. In the same way, a graph has a closed Euler trail if and only if there are as many arcs entering as leaving a vertex. Use the proof of theorem 2.1 to find an algorithm that gives an Euler trail.

2.3 Show that the results given in section 2.1 extend to unoriented graphs provided that $\mathbf{Z}^S$ is replaced by $(\mathbf{Z}/2\mathbf{Z})^S$. In particular, what is the condition that a closed Euler trail exists? Also, give an algorithm for finding a closed Euler trail.

2.4 Show that in a tree $T = (S, A, r)$, the mapping $a \mapsto \omega(a)$ is a bijection between A and $S\backslash r$.

2.5 An intuitive proof of König's lemma: suppose that, at the end of time, the total number of men who have lived on the Earth is infinite, and, suppose, furthermore, that each man had Adam as an ancestor. Then there exists an infinite lineage starting with Adam.

2.6 Show that if $\mathbf{N}^k$ is equipped with the lexicographic ordering, the set $\{x \in \mathbf{N}^k | x \le a\}$ is not, in general, finite. Compare this with propostion 2.7.

2.7 Show that the alphabetic ordering over a free monoid is Artin.

CHAPTER THREE

Rational Subsets of a Monoid and Semi-linear Sets

In this chapter, we take up the problem of effectively manipulating rational and semi-linear subsets. In section 3.1, we introduce these concepts, and in the next section, we show that the set of paths through a graph is rational (this is essentially one of the implications of Kleene's theorem for finite automata).

In section 3.3, semi-linear sets are defined, and it will be shown that they are closed under intersection. This result requires the effective integer solution of linear equations with integer coefficients. This is done in the following section: the proof uses a result from section 3.5 which shows that one can find the set of minimal points for a recursive ideal in $\mathbf{N}^k$.

3.1 Rational Subsets of a Monoid

A *monoid* is a set M, equipped with an associative composition rule and a neutral element for that rule. We write the rule as multiplication:

$(m, n) \mapsto mn$

$M \times M \to M$

The neutral element will be denoted by 1. A monoid M is said to be *commutative* if $mn = nm$, for all $m, n \in M$. In this case, it will then be convenient to denote the rule by addition:

$(m, n) \mapsto m + n$

and to denote the neutral element by 0.

The monoids we need are:
- The free monoid over an alphabet (to be defined in the next section).
- The commutative monoid $\mathbf{Z}^k$ of k-tuples of integers, with the usual (coordinatewise) vector addition – its neutral element is the k-tuple of 0s.
- The commutative monoid $\mathbf{N}^k$ of k-tuples of non-negative integers with addition defined as for $\mathbf{Z}^k$.

A submonoid M' of a monoid M is a subset of M closed under products (i.e., $m, n \in M'$ implies $mn \in M'$) and which contains the neutral element. For example, $\mathbf{N}^k$ is a submonoid of $\mathbf{Z}^k$. It is clear that the intersection of a family of submonoids of M is itself a submonoid of M.

We observe that, for a subset A of M,

$$A^* = \bigcap_{A \subseteq M'} M'$$

where intersection is extended to all the submonoids M' of M containing A. It is clear that A^* is the smallest submonoid of M containing A. We call A^* the *submonoid generated by* A.

Multiplication in M extends to subsets of M. If A and B are two subsets of M, we see that:

$$AB = \{ab \mid a \in A, b \in B\}$$

This operation over subsets of M is associative, distributive with respect to unions, and has a neutral element, the subset $\{1\}$ of M.

We see that:

$$A^n = \underbrace{AA \ldots A}_{n \text{ times}}$$

In particular, $A^0 = \{1\}$, $A^1 = A$. It is easily verified that:

$$A^* = \bigcup_{n \geq 0} A^n$$

In other words, the submonoid generated by A consists of all the elements of M obtained by multiplying some number of elements of A (with possible repetitions – see exercise 3.1).

The set of *rational* (or *regular*) subsets of M is defined inductively as follows:
- Each finite subset of M is rational.
- If A and B are rational subsets of M, then $A \cup B$ and AB are also rational subsets of M.
- If A is a rational subset of M, so too is A^*.

In other words, each rational subset of M is obtained from the finite subsets of M by the finite repetition of the operations union, product and generated submonoid. That is, each rational subset of M is represented by a *rational expression*, which is an expression formed from the finite subsets of M using the binary operations union and product, and using the unary operator $*$. For example, if $a, b \in M$, $\{a, b\}^*$ is a rational expression representing the submonoid generated by a and b. The same submonoid is also represented by the rational expression:

$$\{1, a, b\} \cup \{a, b\}\{a, b\}^*\{a, b\}$$

Another example is:

$$(\{b\} \cup \{a\}\{b\}^*\{a\})^*$$

which represents the set of elements of M obtained by repeated multiplication, in some arbitrary order, of the elements b and a, the latter occurring an even number of times.

A *monoid homomorphism* is a mapping $f: M \to M'$ from the monoid M to the monoid M' which preserves products (i.e., $f(m, n) = f(m)f(n)$, for all m and n in M), and maps the neutral element of M onto the neutral element of M'.

The following result, which is effective, will be of use in Chapter Five.

Proposition 3.1. *Let $f: M \to M'$ be a monoid homomorphism. If L is a rational subset of M, then $f(L)$ is a rational subset of M'.*

PROOF. If L is represented by a rational expression E, then $f(L)$ is represented by the rational expression obtained from E by replacing all occurrences of the elements m of M by $f(m)$. This follows from the following identities (since f is a homomorphism):
- $f(A \cup B) = f(A) \cup f(B)$
- $f(AB) = f(A)f(B)$
- $f(A^*) = f(A)^*$

the proofs of which are left to the reader (cf. exercise 3.2). □

A *length* over a monoid M is an additive homomorphism of M into $\mathbf{N}$, generally denoted by:

$$m \mapsto |m|$$

such that $|m| = 0$ if and only if $m = 1$. For example, the monoid $\mathbf{N}^k$ has a length defined by:

$$(a_1, \ldots, a_k) \mapsto a_1 + \cdots + a_k$$

similarly, for the free monoid: each word is associated with its length in the usual way.

We will need the following proposition:

Proposition 3.2. *Let U and V be two subsets of a monoid M. Then the smallest subset X of M such that:*

$$A = UX \cup V \tag{1}$$

is equal to U^*V. If M has a length, and if $1 \notin U$, it is the unique solution to (1).

PROOF. We have $U^*V = UU^*V \cup V$ because:

$$U^* = \bigcup_{n \geq 0} U^n$$

$$= \{1\} \cup \bigcup_{n \geq 1} U^n$$

$$= \{1\} \cup (U. \bigcup_{n \geq 0} U^n)$$

$$= \{1\} \cup UU^*$$

So, U^*V is a solution to (1). But, if X satisfies (1), then:

$$X = UX \cup V$$

$$= U(UX \cup V) \cup V$$

$$= U^2 X \cup UV \cup V$$

and so,

$$X = U^{n+1}X \cup U^n V \cup \ldots \cup UV \cup V \tag{2}$$

So, X contains $U^n V$ for all n, hence it also contains $U^* V$. Therefore, $U^* V$ is the smallest solution to (1).

Suppose, now, that M has a length and that $1 \notin U$. We show that (1) implies $X \subset U^* V$, which will finish the proof. Let x be an element of X and let $n = |x|$. Then x is in the last member of (2): x is not in $U^{n+1} X$, for, otherwise, $x = u_1 \ldots u_{n+1} x' \Rightarrow |x| = \sum |u_i| + |x'| \geq n + 1$, since $|u_i| \geq 1$, since $1 \notin U$. It follows that x is in $U^n V \cup \ldots \cup UV \cup V$, and, therefore, in $U^* V$. □

3.2 Rationality of the Set of Paths in a Graph

We begin by defining a *free monoid*. We start with a set A, called the *alphabet*. A *word* over A is a finite sequence of *letters*, i.e., elements of A. The empty sequence is called the *empty word*, which we denote by 1. The concatenation of two words defines a third. Thus, the set of words over the alphabet A becomes a monoid with the empty word as its neutral element: we call this monoid the *free monoid* over A, and denote it by A^*. Each word has a *length* $|w|$, and the mapping $w \mapsto |w|$ is a homomorphism from A^* to the additive monoid $\mathbf{N}$ such that $|w| = 0$ if and only if $w = 1$.

Now let $G = (S, A)$ be an oriented graph, where S is the set of vertices of the graph, and where A is the set of edges (or arcs). Each non-empty path can be represented as a word in the free monoid A^*: it is merely the sequence of edges in the path. This representation of the path by a word is faithful. This is not the case for empty paths which are all represented by the empty word: there is a loss of information about the vertex we use to define the empty path, but that is not important for our purposes.

In the treatment that follows, each path is represented by a word over the alphabet of edges, in conformity with the above definitions.

Theorem 3.3. *Let $G = (S, A)$ be a finite, oriented graph, and let s and t be two of the vertices of G. The set of paths from s to t is a rational subset of the free monoid A^*.*

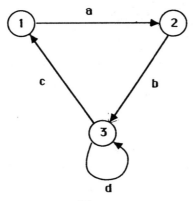

Figure 3.1

The theorem and its proof are illustrated by the following example.

Example 1

In figure 3.1, we denote by C_{ij} the set of paths from i to j. We have $C_{11} = aC_{21} \cup \{1\}$, as can easily be shown. In the same way, $C_{21} = bC_{31}$ and $C_{31} = dC_{31} \cup cC_{11}$. Proposition 3.2, when applied to the last identity, gives $C_{31} = d^* cC_{11}$. By the second identity, we have $C_{21} = bd^* cC_{11}$, and by the first, $C_{11} = \{1\} \cup abd^* cC_{11}$. So, by proposition 3.2, we have

$$C_{11} = (abd^* c)^*$$

which is, therefore, a rational subset of $\{a, b, c, d\}^*$.

∎

The above example shows that it is necessary to solve linear equations, as does the following proposition.

Proposition 3.4. *Let M be a monoid with length, and let A_{ij} $(1 \le i, j \le n)$ and B_i $(1 \le i \le n)$ be two rational subsets of M such that $1 \notin A_{ij}$ for all $i = 1, \ldots, n$, $j = 1, \ldots, n$. Let $X_1, \ldots, X_n$ be subsets of M such that for all $i = 1, \ldots, n$,*

$$X_i = \bigcup_{j=1}^{n} A_{ij} X_j \cup B_i \qquad (3.i)$$

Then each X_i is a rational subset of M.

This result is effective, as the proof shows.

PROOF. Set $U = A_{nn}$, $V = B_n \cup (\bigcup_{j=1}^{n-1} A_n X_j)$. Then equation (3.$n$) can be rewritten as

$$X_n = U X_n \cup V$$

By proposition 3.2, we have (since $1 \notin U$),

$$X_n = U^* V = A_{nn}^* (\bigcup_{j=1}^{n-1} A_{nj} X_j \cup B_n)$$

We now substitute this value for X_n into the equations 3.1 to 3.$n - 1$, obtaining for $i = 1, \ldots, n - 1$:

$$X_i = \bigcup_{j=1}^{n-1} A'_{ij} X_j \cup B'_i \tag{4}$$

with $A'_{ij} = A_{ij} \cup A_{in} A_{nn}^* A_{nj}$ and $B'_i = B_i \cup A_{in} A_{nn}^* B_n$ Since M has a length, we see that $1 \notin A_{ij}$.

By hypothesis on A_{ij} and B_i, the A'_{ij} and B'_i are rational subsets of A^*. By the induction hypothesis on n, (4) shows that each X_i is rational for $i = 1, \ldots, n - 1$. The same is true for U and V, and so, finally, for X_n. □

We finally come to the proof of theorem 3.3, which is a simple consequence of proposition 3.4.

PROOF OF THEOREM 3.3.   We assume that $S = \{1, \ldots, n\}$ and that $s = 1$. Let X_i be the set of paths from i to t. Let A_{ij} be the set of edges from i to j. Each A_{ij} is finite, so is a rational subset of A^*. Let $B_i = \emptyset$ if $i \neq t$, and let $B_t = \{1\}$. We then have for $i = 1, \ldots, n$

$$X_i = A_{ij} X_j \cup B_i$$

whence the theorem, by proposition 3.4. □

3.3 Semi-linear Sets

In this section and the next, we will be concerned with the study of the rational subsets of commutative monoids. The law for these monoids will be denoted by +, and the neutral element by 0. Let M be a commutative monoid. A *linear subset* of M is a subset of the form:

$$u + V^*$$

where u is in M and V is a finite subset of M. In other words, if $V = \{v_1, \ldots, v_p\}$, a linear subset of M is a subset of the form:

$$\{u + \sum_{i=1}^{p} n_i v_i | n_i \in \mathbf{N}\}$$

This is derived from the fact that the submonoid V^* of M generated by V is identical to the set of elements of the form:

$$n_1 v_1 + \cdots + n_p v_p \tag{5}$$

i.e., to the N-linear combinations of elements of V.

A semi-linear subset of M is a finite union of linear subsets of M, i.e:

$$\bigcup_{i=1}^{n} (u_i + V_i^*) \tag{6}$$

To effectively specify a semi-linear subset of M, we specify the elements $u_1, \ldots, u_n$ and the finite subsets $V_1, \ldots, V_n$.

From the definition of the rational subsets of a monoid, it easily follows that each semi-linear subset of M is a rational subset of M. It should be noted that the product AB of two subsets of M has here been replaced by their sum $A + B$ because of our conventions. We now have the following proposition.

Proposition 3.5. *Let M be a commutative monoid. A subset of M is rational if and only if it is semi-linear.*

As the proof shows, the passage from a rational expression representing A to a *semi-linear expression* (i.e., an expression of the same form as (6)) is effective.

PROOF. It is enough to show that the set of semi-linear subsets of M contains the finite subsets of M, and that the set is closed under union, sum and *.

We can put every finite subset of M into the same form as (6) with the V_i being set to $\emptyset$.

Now, if $A = \bigcup_i (u_i + V_i^*)$, and $B = \bigcup_j (v_j + W_j^*)$, then $A \cup B$ is clearly semi-linear. Furthermore, by the distributivity of addition over union:

$$A + B = \bigcup_{ij} (u_i + v_j + V_i^* + W_j^*)$$

But, in a commutative monoid, we have:

$$V^* + W^* = (V \cup W)^* \tag{7}$$

which follows from (5). Consequently, $A + B$ is semi-linear. But, we also have:

$$A^* = (\bigcup_{i=1}^{n} (u_i + V^*))^*$$

$$= \sum_{i=1}^{n} (u_i + V_i^*)^*$$

which follows from (7) by induction. It is then sufficient to show that $(u + V^*)^*$ is semi-linear, since the sum of semi-linear sets is semi-linear, as we have seen. But, we have

$$(u + V^*)^* = (u + (\{u\} \cup V)^*) \cup \{0\}$$

as can be seen (cf. exercise 3.8). Since $\{u\} \cup V$ is finite, $(u + V^*)^*$ is semi-linear. □

We need to consider the cases of the monoids $\mathbf{Z}^k$ and $\mathbf{N}^k$ for the purposes of the present study. We begin by proving the following proposition.

Proposition 3.6.   *Let A be a semi-linear subset of $\mathbf{Z}^k \subset \mathbf{N}^k$. Then A is a semi-linear subset of $\mathbf{N}^k$.*

PROOF. We have $A = \bigcup_i (u_i + V_i^*)$, where $u_i \in \mathbf{Z}^k$, and where V_i is a finite subset of $\mathbf{Z}^k$. But, $A \subset \mathbf{N}^k$, so $u_i \in A \Rightarrow u_i \in \mathbf{N}^k$. Furthermore, let $v_i \in V_i$. Then, for every positive integer n, we have $u_i + nv_i \in A$. Then, $u_i + nv_i \in \mathbf{N}^k$, which is only possible if $v_i \in \mathbf{N}^k$. □

The following result is important. We will need a theorem about linear equations (which is proved in the next section) in order to prove it.

Theorem 3.7. *The intersection of two semi-linear subsets of $\mathbf{Z}^k$ is a semi-linear subset of $\mathbf{Z}^k$.*

Again, all constructions are effective.

PROOF. By the distributivity of intersection over union, it will be enough to prove this theorem for linear subsets of $\mathbf{Z}^k$. Therefore, let $A = u + U^*$, and let $B = v + V^*$, where $u, v \in \mathbf{Z}^k$, and where U and V are finite subsets of $\mathbf{Z}^k$. With $a = |U|$ and $b = |V|$, there exist monoid homomorphisms $\alpha : \mathbf{N}^a \to \mathbf{Z}^k$ and $\beta : \mathbf{N}^b \to \mathbf{Z}^k$ such that $\alpha(\mathbf{N}^a) = U^*$ and $\beta(\mathbf{N}^b) = V^*$. Indeed, if $U = \{u_1, \ldots, u_a\}$, it is enough to define $\alpha(n_1, n_2, \ldots, n_a) = \sum_{i=1}^{a} n_i u_i$ and similarly for β.

Let ϕ and τ be the homomorphisms:

$$\phi : \mathbf{N}^a \times \mathbf{N}^b \to \mathbf{Z}^k$$

$$\phi(x, y) = \alpha(x) - \beta(y)$$

$$\tau : \mathbf{N}^a \times \mathbf{N}^b \to \mathbf{Z}^k$$

$$\tau(x, y) = \alpha(x)$$

Let $W = \{(x, y) \in \mathbf{N}^a \times \mathbf{N}^b | \phi(x, y) = v - u\}$. Using theorem 3.9 (which will be proved in the next section), W is a semi-linear subset of $\mathbf{Z}^k$. It follows that $\tau(W)$ is a semi-linear subset of $\mathbf{Z}^k$, as it is the image under a homomorphism of a semi-linear subset (cf. propositions 3.1 and 3.5). Finally, $u + \tau(W)$ is semi-linear.

We show that $u + \tau(W) = A \cap B$. Indeed, if $w \in u + \tau(W)$, then $w = u + \tau(x, y)$, with $\phi(x, y) = v - u$. So, $w = u + \alpha(x) \in u + U^* = A$, and $\alpha(x) - \beta(y) = \phi(x, y) = v - u \Rightarrow w = u + \alpha(x) = v + \beta(y) \in v + V^*$, from which we obtain $w \in A \cap B$. Conversely, if $w \in A \cap B$, then $w = u + \alpha(x) = v + \beta(y)$, with $x \in \mathbf{N}^a$, $y \in \mathbf{N}^b$. Then, $\phi(x, y) = \alpha(x) - \beta(y) = v - u$, so $(x, y) \in W$, and $w = u + \tau(x, y) \in u + \tau(W)$. □

If J is a subset of $\{1,\ldots,k\}$, we will use x_J to denote the projection of the vector $x \in \mathbf{N}^k$ onto $\mathbf{N}^J$. The easy proposition which follows will be of use in Chapter Five.

Proposition 3.8.   *Let L be a semi-linear subset of $\mathbf{N}^K$, with $L = \bigcup_i(w_i + V_i^*)$. Let J be a subset of $\{1,\ldots,K\}$.*
>   1. *The following two properties are equivalent:*
>   *(i) For each integer, N, there exists an element x of L such that $x_J \geq (N,\ldots,N)$.*
>   *(ii) There exists an i such that $(\sum_{v \in V_i} v_J) \geq (1,\ldots,1)$.*
>   *In particular, this property is decidable.*
>   2. *If the projection from L onto $N^{\{1,\ldots,K\}\setminus J}$ is a singleton, and if L has either property (i) or (ii), then L has a subset of the form $w + Nt$, with $\forall j \in \{1,\ldots,K\}: j \in J \Leftrightarrow t_j \geq 1$.*

PROOF. 1.   That (ii) implies (i) is immediate, since L contains $x = w_i + n(\sum_{v \in V_i} v)$ for all integers $n \geq 0$. Let us assume that (ii) is not true: $\forall i, \exists j \in J$ such that $\forall t \in V_i, t_j = 0$. So, if $x \in w_i + V_i^*$, we have $x_j = (w_i)_j$. But every x in L is in one of the $w_i + V_i^*$. One cannot have (i). Therefore, (i) and (ii) are equivalent. Now, it is clear that property (ii) is decidable.

2.   Under these assumptions, we have: $\forall j \in \{1,\ldots,K\}\setminus J, \forall i, \forall s \in V_i, s_j = 0$. There exists an i such that $(\sum_{s \in V_i})_J \geq (1,\ldots,1)$. Let $t = \sum_{s \in V_i} s$, then L contains $w_i + Nt$, and we have $j \in J \Rightarrow t_j \geq 1$ and $j \notin J \Rightarrow t_j = 0$. □

3.4 Natural Solutions to Systems of Linear Equations

We begin by stating the principal result of this section. Immediately thereafter, we will formulate it in terms of linear equations.

Theorem 3.9.   *Let $\phi: \mathbf{N}^k \to \mathbf{Z}^m$ be a monoid homomorphism, and let c be in $\mathbf{Z}^m$. Then the set:*

$$\phi^{-1}(c) = \{x \in \mathbf{N}^k \,|\, \phi(x) = c\}$$

is a semi-linear subset of $\mathbf{N}^k$ which can be effectively constructed.

A homomorphism $\phi: \mathbf{N}^k \to \mathbf{Z}^m$ is completely defined by its effect on the canonical base of $\mathbf{N}^k$, i.e., the elements $e_1, \ldots, e_k$, with

$$e_i = (0, \ldots, 0, 1, 0, \ldots, 0)$$

$$\uparrow$$

i^{th} position

Set $\phi(e_j) = (a_{ij})$, $1 \le i \le m$, $c = (c_i)$, $1 \le i \le m$, and $x = (x_j)$, $1 \le j \le k \in \mathbf{N}^k$. Then the equation $\phi(x) = c$ is equivalent to the following system of linear equations:

$$a_{i1} x_1 + a_{i2} x_2 + \cdots + a_{ik} x_k = c_i, \; 1 \le i \le m \tag{8}$$

In other words, theorem 3.9 states that *the set of k-tuples of natural numbers which are solutions of a system of linear equations with integer coefficients is a semi-linear subset of* $\mathbf{N}^k$. This set is certainly effectively computable, as we shall see.

The proof of theorem 3.9 is easy if we disregard the question of effectiveness. Recall, first, that the natural ordering over $\mathbf{N}^k$ is defined by:

$$x \le y \Leftrightarrow \forall i \in \{1, \ldots, k\}, \; x_i \le y_i$$

Let A be a subset of $\mathbf{N}^k$. The set of minimal elements of A, i.e.:

$$\min(A) = \{x \in A \,|\, \forall y \in A, y \le x \Rightarrow y = x\}$$
$$= \{x \in A \,|\, \forall y \in A, \; \text{not}(y < x)\}$$

is obviously formed of elements that are pairwise incomparable. By theorem 2.5(ii), this set is, therefore, finite. We define the following two finite sets:

$$U = \min\{x \in \mathbf{N}^k \,|\, \phi(x) = c\} = \min(\phi^{-1}(c))$$

$$V = \min\{x \in \mathbf{N}^k \,|\, x \ne 0, \phi(x) = 0\} = \min(\phi^{-1}(0) \backslash 0)$$

We shall show that $\phi^{-1}(c) = U + V^*$, which implies that $\phi^{-1}(c)$ is semi-linear. We show first that:

$$\phi^{-1}(0) = V^*$$

Indeed, it is clear that $V^* \subset \phi^{-1}(0)$, since, for each x in V, we have $\phi(x) = 0$.

Now, if $x \in \phi^{-1}(0)$, we have either that $x = 0$ or that $x \in V$, and so x is in V^* or we have that $x \notin V \cup \{0\}$. Then there exists $y < x$, $y \in \phi^{-1}(0)$, $y \neq 0$. So, we have y, $x - y \in \phi^{-1}(0)$, and these two vectors each have a size (the sum of the coordinates of a vector) strictly less than the size of x. By induction, we obtain $y, x - y \in V^*$, from which we obtain:

$$x = y + (x - y) \in V^*$$

To show that $\phi^{-1}(c) = U + V^*$, a similar argument is employed. First, $U + V^* \subset \phi^{-1}(c)$. Conversely, let $x \in \phi^{-1}(c)$. If $x \notin U$, there exists $y \in \phi^{-1}(c)$ with $y < x$. Then, $y \in U + V^*$ by induction, and $x - y \in \phi^{-1}(0)$, so $x - y \in V^*$ by the preceding argument. So, $x = y + (x - y) \in (U + V^*) + V^* \subset U + V^*$.

To complete the proof of theorem 3.9, it is enough to show that we can effectively compute the sets U and V required above. In other words, given a monoid homomorphism $\phi: \mathbf{N}^k \rightarrow \mathbf{Z}^k$ and a vector c in $\mathbf{Z}^k$, it is a matter of finding $\min(\phi^{-1}(c))$ and $\min(\phi^{-1}(0)\backslash 0)$. It is enough to be able to compute $\min(\phi^{-1}(c)\backslash 0)$, since $c \neq 0 \Rightarrow \phi^{-1}(c)\backslash 0 = \phi^{-1}(c)$.

To compute $\min(\phi^{-1}(c)\backslash 0)$, we borrow notations and results from the next section. Thus, let I be the ideal of $\mathbf{N}^k$ generated by $\phi^{-1}(c)\backslash 0$, i.e., $I = \{x \in \mathbf{N}^k | \exists y \in \phi^{-1}(c)\backslash 0, x \geq y\}$. We have $\min(I) = \min(\phi^{-1}(c)\backslash 0)$.

Using theorem 3.12 (next section), it is enough to show that $\overline{I}$ is a recursive ideal of $\mathcal{N}^k$, where $\overline{I}$ is defined by $\overline{I} = \{x \in \mathcal{N}^k | \exists y \in \mathbf{N}^k\backslash 0, \phi(y) = c, \text{ and } y \leq x\}$. It is enough, then, to be able to decide whether, given a in $\mathcal{N}^k$, we have $a \in \overline{I}$. But $a \in \overline{I}$ is equivalent to the system:

$$\begin{cases} \phi(y) = c \\ \forall i \in \{i, \ldots, k\}, a_i \neq \infty \Rightarrow y_i \leq a_i \\ y \neq 0 \end{cases}$$

We will define a finite number of systems of linear equations such that the above system has a solution if and only if one of these systems has a non-zero solution. This will reduce the problem to one of showing that we can decide whether a system of linear equations has a non-zero solution. These systems are indexed by the finite set:

$$B = \{b \in \mathcal{N}^k \mid a_i \neq \infty \Rightarrow b_i \leq a_i, a_i = \infty \Rightarrow b_i = \infty\}$$

The system corresponding to B is defined as follows:

$$\begin{cases} \phi(y) = c \\ y_i = b_i (i = 1, \ldots, k, a_i \neq \infty) \end{cases}$$

As has already been pointed out, this reduces to showing that one can decide whether a system of linear equations has a non-zero solution. We will write such a system as:

$$y_1 u_1 + \cdots + y_k u_j = c \tag{9}$$

where the u and c are in $\mathbf{Z}^m$. It is enough to prove:

Lemma 3.10. *We can decide whether (9) has a non-zero natural solution.*

PROOF. (By induction on k.) If $k = 1$, the result is immediate. Assume, then, that $k \geq 2$. If the vectors $u_1, \ldots, u_k$ are linearly independent, the lemma follows from the theory of linear equations: in this case, there exists at most one vector $(x_1, \ldots, x_k) \in \mathbf{Q}^k$ satisfying (9) and which is effectively computable: it is now sufficient to show that the vector has natural number coefficients that are not all zero. We can thus assume that the vectors u_i are linearly independent. There then exist effectively computable natural numbers $a_1, \ldots, a_k$, and disjoint subsets I and J of $\{1, \ldots, k\}$ such that $I \cup J = \{1, \ldots, k\}$, such that:

$$\text{There is an } i \text{ in } I \text{ with } a_i \neq 0 \tag{10}$$

and such that:

$$\sum_{i \in I} a_i u_i - \sum_{j \in J} a_j u_j = 0 \tag{11}$$

We show, if there is a non-zero solution (x_i) of (9), then there exists one satisfying:

$$\exists i \in I, x_i \leq a_i$$

Indeed, if there is a non-zero solution (x_i) of (9) such that:

$$\forall i \in I, x_i > a_i \tag{12}$$

then, by (11) and (9),

$$\sum_{i \in I}(x_i - a_i)u_i + \sum_{j \in J}(x_j + a_j)u_j = c$$

so $y \in \mathbf{N}^k$ defined by:

$$y_i = x_i - a_i, \text{ if } i \in I$$
$$y_j = x_j + a_j, \text{ if } j \in J$$

is a solution of (9), which is non-zero by (12), and which satisfies:

$$\forall i \in I, y_i \leq x_i$$

and, by (10):

$$\exists i \in I, y_i < x_i$$

It suffices to replace x by y and to iterate this procedure to arrive at a non-zero solution of (9) such that $\exists i \in I, x_i \leq a_i$.

For each $i_0 \in I$, and for each h, $0 \leq h \leq a_{i_0}$, we now define the system:

$$\sum_{i \neq i_0} x_i u_i = c - h u_{i_0} \tag{i_0, h}$$

By induction, we can decide whether this system has a non-zero solution.

The above argument shows that (9) has a non-zero solution if and only if at least one of the systems (i_0, h) has a non-zero solution, or if one of the systems $(i), h)$, with $h \geq 1$, has the non-zero solution. Since the number of systems (i_0, h) is finite, the proof is complete. $\square$

3.5 Recursive Ideals

An *ideal* in a monoid M is a subset I of M such that for all $x \in I$, $y \in M$, we have:

$$xy, yx \in I$$

In the monoids $\mathbf{N}^k$ and $\mathcal{N}^k = (\mathbf{N} \cup \{\infty\})^k$ equipped with componentwise addition, an ideal is simply characterized by:

$$x \in I \Rightarrow x + y \in I$$

for all y in the monoid, or by:

$$x \in I, y \geq x \Rightarrow y \in I$$

where $\geq$ is the natural ordering.

The intersection of a family of ideals is again an ideal. The *ideal generated* by a subset A is:

$$\bigcap_{A \subseteq I} I$$

where intersection is extended to all the ideals containing A. This ideal is also equal to

$$A + M = \{x \in M \mid \exists y \in M, \exists z \in A, x = y + z\}$$

as is easily shown (cf. exercise 3.14). When $M = \mathbf{N}^k$ or $M = \mathcal{N}^k$, we therefore also have that the ideal generated by a subset A is:

$$\{x \in M \mid \exists y \in A, x \geq y\}$$

Let I be an ideal of $\mathbf{N}^k$. We use $\bar{I}$ to denote the ideal of $\mathcal{N}^k$ generated by I in $\mathcal{N}^k$.

Lemma 3.11. *Let I be an ideal of $\mathbf{N}^k$. Then $I = \bar{I} \cap \mathbf{N}^k$.*

PROOF. We have $I \subset \bar{I}$ and $I \subset \mathbf{N}^k$, so $I \subset \bar{I} \cap \mathbf{N}^k$. Now, let x be in $\bar{I} \cap \mathbf{N}^k$. We have seen that:

$$\bar{I} = \{x \in \mathcal{N}^k \mid \exists y \in I, x \geq y\}$$

Hence, there then exists a y in I such that $x \geq y$. Since I is an ideal, we infer, then, that $x \in I$. □

We will say that an ideal I of $\mathcal{N}^k$ is *recursive* if there exists an algorithm which allows us to decide whether an element of $\mathcal{N}^k$ is in I.

Theorem 3.12.   *Let I be an ideal of* $\mathbf{N}^k$.
  (i) *The set of minimal elements, $\min(I)$, of I, is finite and generates I. We also have $a\colon I = \min(I) + \mathbf{N}^k$. In particular, I is semi-linear.*
  (ii) *If $\overline{I}$ is a recursive ideal of $\mathcal{N}^k$, then $\min(I)$ is effectively computable and I is effectively semi-linear.*

The converse of (ii) is true (and easier to prove): if $\min(I)$ is known, then $\overline{I}$ is a recursive ideal of $\mathcal{N}^k$ (cf. exercise 3.17). We point out, in order that the theorem be better understood, that if I is an ideal of $\mathbf{N}$ for which we know how to test membership, we can only compute $\min(I)$ if we know how to decide whether $I = \emptyset$: this is equivalent to $\infty \in \overline{I}$.

Example 2

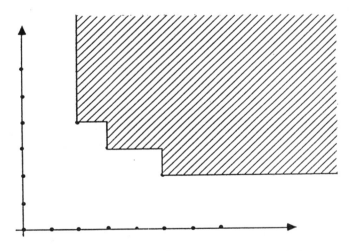

Figure 3.2 An ideal of $\mathbf{N}^2$ and its three minimal elements

■

PROOF OF THEOREM 3.12.   (i) The set $\min(I)$ is finite by theorem 2.5(ii). $\min(I)$ generates I because, for each element x of I, either $x \in \min(I)$ or $x \notin \min(I)$, and in this case there exists $x' \in I$ such that $x' < x$.

By induction, there exists $y \in \min(I)$ such that $y \leq x$. Thus, every element of I is greater than or equal to some element of $\min(I)$, so $\min(I)$ generates I. The remainder of (i) is obvious.

(ii) It suffices to show that $\min(I)$ is effectively computable. To do this, it is enough to show that we can find a bound b of all the coordinates of the elements of $\min(I)$. Then we will have:

$$\min(I) \subset \{x \in \mathbf{N}^k \mid x \leq (b, \ldots, b)\}$$

and, since the set on the right-hand side is finite, we can effectively find all the minimal elements of I, since we can test membership of $\overline{I}$, and therefore membership of I (lemma 3.11).

We first show that b is such a bound if and only if we have:

$$\forall x \leq (b, \ldots, b), \ \overline{x} \in \overline{I} \Rightarrow x \in I \tag{13}$$

where $\overline{x}$ is the element of $\mathcal{N}^k$ defined by $\overline{x}_i = x_i$ if $x_i < b$, $\overline{x}_i = \infty$ if $x_i = b$.

Let b, therefore, be a bound of all the coordinates of the elements of $\min(I)$, and let $x \leq (b, \ldots, b)$ such that $\overline{x} \in \overline{I}$. The last condition implies that there exists $a \in \min(I)$ such that $a \leq \overline{x}$ (because I is generated by $\min(I)$ – the same applies to $\overline{I}$ as an ideal of $\mathcal{N}^k$). For each coordinate i, we have either $x_i < b$, so $\overline{x}_i = x_i$ or $x_i = b$, so, since b is a bound $a_i \leq b = x_i$. In each case, $a_i \leq x_i$, from which we obtain $a \leq x$ and x is in I.

Conversely, let b be such that (13) is satisfied, and $a \in \min(I)$. We show that $a \leq (b, \ldots, b)$. In the opposite case, let $x = \inf(a, (b, \ldots, b))$, so $x < a$ and $x \notin I$. We also have that $x \leq (b, \ldots, b)$; furthermore, for all coordinates i, we have either $a_i < b$ and so $x_i = a_i$ and $\overline{x}_i = a_i$; or $a_i \geq b$, and so $x_i = b$ and $\overline{x}_i = \infty \geq a_i$. In each case, $\overline{x}_i \geq a_i \Rightarrow \overline{x} \geq a \Rightarrow \overline{x} \in \overline{I}$. This contradicts (13).

To finish the proof, we observe that given b, condition (13) is decidable, since the set $\{x \mid x \leq (b, \ldots, b)\}$ is finite, and that we know how to test whether a vector is in $\overline{I}$ (or in I by lemma 3.11). To find a bound b, we verify (13) for $b = 0, 1, 2, \ldots$. The process must terminate because $\min(I)$ is finite and, therefore, a bound exists. $\square$

Exercises

3.1 Show that if A is a subset of a monoid, then the submonoid generated by A is $\bigcup_{n \in N} A^n$.

3.2 Show that if $f: M \to M'$ is a monoid homomorphism, then for each
submonoid P of M (respectively of M'), $f(P)$ (respectively, $f^{-1}(P)$)
is a submonoid of M' (respectively of M). Show that $f(A^*) = f(A)^*$
for every subset A of M. Find an example where $f^{-1}(B)^* \neq f^{-1}(B^*)$.

3.3 Show that if the monoid M has a length, the monoid formed from
the finite subsets of M also has one.

3.4 Show that if R_i is a rational subset of $M_i (i = 1, 2)$, then $R_1 \times R_2$ is
a rational subset of $M_1 \times M_2$.

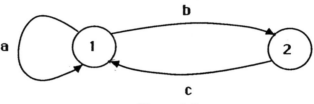

Figure 3.3

3.5 Give a rational expression for the set of paths $1 \to 1$ shown in figure
3.3. Generalization: show that if G is a graph with a vertex s such
that each closed path passes through s, the set of paths $s \to s$ is of
the form C^* for a particular finite set C of paths in G.

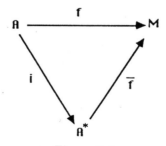

Figure 3.4

3.6 Justification of the term *free monoid*. Show that for every monoid M, and for all functions $f: A \to M$, there exists exactly one extension of f to a homomorphism $\bar{f}$ from the monoid A^* to M such that diagram in figure 3.4 is commutative, where i is the natural injective function $A \to A^*$. What is the analogous property for $\mathbf{N}^k$?

3.7 Let A be a semi-linear subset of $\mathbf{N}^k$. Show that for x in $\mathbf{N}^k$, we can decide whether x is in A.

3.8 Let $L = a + B^*$ be a linear subset of a commutative monoid M. Show that $L^* = \{0\} \cup (a + (\{a\} \cup B)^*)$. Note that if $M = \mathbf{N}^2$, $a = (1,0)$, and $B = \{(0,1)\}$, the identity reduces to $\{(x,y) \in \mathbf{N}^2 | x = 1\}^* = \{(0,0)\} \cup \{(x,y) \in \mathbf{N}^2 | x \geq 1\}$. The general case can be deduced by substitution.

3.9 Show that if M is a *subtractive* monoid of $\mathbf{N}^k$ (i.e., $x, x + y \in M \Rightarrow y \in M$), then M is generated by the minimal elements of $M \backslash \{0\}$. Show that M is a linear subset of $\mathbf{N}^k$ using theorem 2.5(ii). Show that a submonoid M of $\mathbf{N}^k$ is subtractive if and only if there exists a subgroup H of $\mathbf{Z}^k$ such that $M = H \cap \mathbf{N}^k$.

3.10 Show that $\{(x,y) \in \mathbf{N}^2 | x = y = 0$ or $x \geq 1\}$ is a submonoid of $\mathbf{N}^2$ which is not finitely generated, but is, nevertheless, semi-linear.

3.11 Show that $\{(x,y) \in \mathbf{N}^2 | y \leq x^2\}$ is not a semi-linear subset of $\mathbf{N}^2$.

3.12 Deduce from theorem 3.9 that the set of vectors in $\mathbf{N}^k$ which are solutions to a system of linear inequalities with coefficients in $\mathbf{Z}$ is a semi-linear subset of $\mathbf{N}^k$ that is effectively computable.

3.13 Find a method to solve in $\mathbf{Z}$ a system of linear equations with coefficients in $\mathbf{Z}$. Hints: look for a non-zero coefficient with the least value, say $a_1 > 0$, which appears in the equation $a_1 x_1 + \cdots + a_k x_k = c$. If $a_1 = 1$, eliminate x_1 from the other equations. If $a_1 > 1$, and if the other a_i are divisible by a_1, either c is not divisible by a_1, and there is indeed no solution, or c is divisible by a_1, and one can reduce the size of the system. If $a_1 > 1$, and if there is an a_i that is not divisible by a_1, let $a_i = a_1 q_i + r_i$ (euclidean division by a_1). Replace the above equation by $a_1 x_1' + r_2 x_2 + \cdots + r_k x_k$, where $x_1' = x_1 + q_2 x_2 + \cdots + q_k x_k$, which allows the elimination of x_1 from the other equations (see Knuth, 1981, 4.5.2).

3.14 Show that if A is a subset of an additive monoid M, then the ideal generated by A is $A + M$.

3.15 Show that if M is a group, every non-empty ideal of M is identical to M.

3.16 Show that if I is an ideal of $\mathbf{N}^k$, $I = \emptyset$ if and only if $(\infty, \ldots, \infty) \in \overline{I}$. Show that if $\overline{I}$ is a recursive ideal, we can decide whether $I = \emptyset$.

3.17 Let I be an ideal of $\mathbf{N}^k$. Show that $x \in \mathcal{N}^k$ is in $\overline{I}$ if and only if there exists $a \in \min(I)$ such that $x \geq a$. In particular, if $\min(I)$ is known, $\overline{I}$ is a recursive ideal.

3.18 Show that if $I \subset \mathbf{N}^k$ is an ideal whose minimal elements are known, the complement of I in $\mathbf{N}^k$ is an effectively computable semi-linear set.

Notes

Theorem 3.1 is a particular case of Kleene's theorem (more precisely, the assertion 'recognizable' implies 'rational'): a language is recognizable by a finite-state automaton if and only if it is rational (or regular) – see, for example, Eilenberg (1974, theorem VIII.5.1). Theorem 3.7 is by Ginsburg and Spanier (1964), as is theorem 3.9. Our presentation here was inspired jointly by Ginsburg (1966) and by Eilenberg and Schützenberger (1969), and makes use of theorem 3.12 which is due to Hack (1974, 1979; see also Jantzen and Valk, 1980, Valk and Jantzen, 1985). An important result, which is not given here, is the closure by complementation of semi-linear subsets, due to Ginsburg and Spanier (1966), of which one very special case is given in exercise 3.18; see also Eilenberg and Schützenberger (1969). For algorithmic complexity problems in semi-linear sets, Presburger arithmetic and natural-value linear equations, see Fischer and Rabin (1974), von zur Gathen and Sieveking (1978), Huet (1978), Mayr and Meyer (1982), Huynh (1982, 1985, 1985a).

CHAPTER FOUR

Vector Addition Systems
with States

With this chapter we reach the heart of the matter. The first section presents the concepts of vector addition systems with states, walks and admissible paths. The second section is longer: it contains the definition of Karp and Miller trees and the proofs of some of their properties. On the one hand, these trees are finite, on the other they provide ways of regulating the infinite behavior of a system.

Covering graphs are presented in section 4.3. They permit the extension of each walk in a VASS to a path in the covering graph: we infer the effective extensions from those with coordinates that are not potentially infinite. The last section is devoted to the proof of a technical result which assures us that, under certain conditions, positive walks exist.

4.1 Definitions

A *vector addition system with states* (VASS) is a finite oriented graph $G = (Q, A)$, an integer $m \geq 1$ and a mapping $v : A \to \mathbf{Z}^m$.

Example 1

In figure 4.1, $m = 3$. For convenience, we show the vector $v(x)$ on each arc x.
∎

In other words, a VASS is a graph labeled in $\mathbf{Z}^m$. The *valuation* (or *label*) of an arc a is $v(a)$.

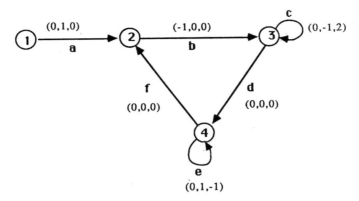

Figure 4.1

A vertex of G will also be called a *state*, and an arc will be called a *transition*. If c is a path in G, it valuation is, by definition, the sum of the valuations of the arcs making up c:

$$c = a_1 \cdots a_n \Rightarrow v(c) = \sum_{i=1}^{n} v(a_i)$$

Note that if c is an empty path, we have $v(c) = 0$, i.e., the vector $(0, \ldots, 0)$ in $\mathbf{Z}^m$.

A *configuration* of a VASS is a pair (p, x) where p is a state of G, and x is a vector in $\mathbf{Z}^m$. We call p the *state* of the configuration, and x the *vector* or the *point* of the configuration.

A *walk* in G is a sequence of configurations determined by a path in G. More formally, a walk is a pair (c, x) formed from a path c and a vector x in $\mathbf{Z}^m$. If the path is $c = a_1 \cdots a_n$, then the sequence of configurations in the walk is:

$$(p_0, x_0), (p_1, x_1), \ldots, (p_n, x_n)$$

with $x_0 = x$, $x_i = x_{i-1} + v(a_i)$, for $i = 1, \ldots, n$, $p_0 = \alpha(a_1)$ and $p_i = \omega(a_i)$ for $i = 1, \ldots, n$ (recall that $\alpha(a)$ and $\omega(a)$ are the start and end of an arc a). Each of the configurations (p_i, x_i) will be called an *intermediate configuration* of the walk. Furthermore, p_i will be called an *intermediate state*, and x_i an *intermediate point* in the walk. For $i = 0$ (respectively, $i = n$), they will be called the *initial configuration*, the *initial state* and the *initial point* (respectively, the *final configuration*, the *final state*, and the *final point*) of the walk.

In example 1, let P be the walk $P(abcdeef, (2, 0, 0))$. The successive intermediate configurations of P are:

$$(1, (2, 0, 0)); \quad (2, (2, 1, 0)); \quad (3, (1, 1, 0));$$

$$(3, (1, 0, 2)); \quad (4, (1, 0, 2)); \quad (4, (1, 1, 1));$$

$$(4, (1, 2, 0)); \quad (2, (1, 2, 0))$$

We will say that the walk (c, x) goes from (p, x) to (q, y) if these two configurations are the initial and final configurations, respectively. This will be denoted by:

$$(p, x) \xrightarrow{c} (q, y) \tag{1}$$

The notation:

$$(p, x) \rightarrow (q, y)$$

will denote the fact that there exists a walk (c, x) for which (p, x) is the initial, and (q, x) the final, configuration. In other words, this notation will be used to represent the fact that there exists a path c in a VASS such that $\alpha(c) = p$, $\omega(c) = q$, and $y = x + v(c)$. We will say in this case that the configuration (q, y) is *accessible* from the configuration (p, x). A walk (c, x) is said to be *positive* if all of its intermediate points are in $\mathbf{N}^m$: this we denote by:

$$(p, x) \xRightarrow{c} (q, y)$$

We will say in this case that the path c is *admissible for the point* x.

An obvious, but fundamentally important, remark is the following: if a path is admissible for a point x, then it is also admissible for all points $\geq x$ (under the natural ordering over $\mathbf{N}^m$).

As above, the notation:

$$(p, x) \Rightarrow (q, y)$$

will denote the existence of a path $c: p \rightarrow q$ such that $y = x + v(c)$ and such that the walk (c, x) is positive. We will then say that the configuration (q, y) is *positively accessible* from the configuration (p, x).

The *accessibility problem* for a VASS is to decide whether, given two configurations, one is positively accessible from the other – this problem will be solved in the next chapter within a slightly more general framework.

In view of these future applications, we must slightly generalize all of the preceding definitions. Let J be a subset of $\{1, \ldots, m\}$. We say that a walk (c, x) is *positive with respect to J* if, for all intermediate points z, $z_J \geq 0$ (recall that z_J is the projection of z onto $\mathbf{N}^J$). We will use the notation:

$$(p, x) \overset{c}{\Rightarrow}_J (q, y)$$

to denote the fact that the walk (c, x) is positive with respect to J with final configuration (q, y), and we will write:

$$(p, x) \Rightarrow_J (q, y)$$

to denote the fact that there exists a walk (c, x) which is positive with respect to J and has final configuration (q, y).

We will also say that a path c is *admissible for x with respect to J*. We observe that, as before, if a path is admissible for a point x with respect to J, then it is also admissible with respect to J for all points y such that $y_J \geq x_J$ under the natural odering on $\mathbf{N}^J$.

4.2 Karp and Miller Tree

We adjoin a new element, denoted by ∞, to $\mathbf{Z}$. The ordering and the addition operator are extended to $\mathbf{Z} \cup \infty$ in the obvious way:

$$\forall n \in \mathbf{Z} \cup \infty, n + \infty = \infty, \text{ and } n \leq \infty$$

Consider a VASS whose underlying graph is $G = (Q, A)$ whose valuation is v, and let $(p, x) \in Q \times \mathbf{N}^m$ be an *initial configuration*.

A *generalized configuration* is an element of $\mathbf{Q} \times (\mathbf{Z} \cup \infty)^m$. If $x \in (\mathbf{Z} \cup \infty)^m$ and $j \in \{1, \ldots, m\}$, j will be called a *finite coordinate* of x if $x_j \neq \infty$, and an *infinite coordinate* of x if $x_j = \infty$.

The *Karp and Miller tree* associated with a VASS and with the initial configuration (p, x) is a tree T in which each vertex is labeled with a generalized configuration, and in which each arc is labeled with an element of A (i.e., is an arc of G). The tree T is defined recursively by the following rules:

    I. The root ρ of T is labeled (p, x).

    II. Let σ be a vertex already in T and labeled (q, y), then:

        (a) If the label of σ is identical to the label of one of the ancestors of σ, then σ has no sons.

        (b) In the alternative case, σ has a son, $\sigma(a)$, for each arc $a \in A$, starting at q in G, and such that $y + v(a) \geq 0$. The arc $\sigma \to \sigma(a)$ in T is labeled by a, and $\sigma(a)$ is labeled by $(\omega(a), z)$, where z_i is defined for each coordinate $i \in \{1, \ldots, m\}$ by: (i) if there is an ancestor of $\sigma(a)$ in T whose label is $(\omega(a), z')$ with $z' \leq y + v(a)$ and $z_i' < (y + v(a))_i$, then $z_i = \infty$; (ii) otherwise, $z_i = (y + v(a))_i$.

Example 2

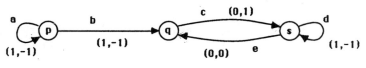

Figure 4.2

We construct the tree associated with the VASS shown in figure 4.2: its initial configuration is $(p, (1, 1))$ (we write this simply as $p, 1, 1$). Note that each arc is annotated with its valuation. The tree is that shown in figure 4.3.

∎

We now present the two principal results of this section.

Theorem 4.1. *A Karp and Miller tree is finite. In particular, it can be effectively constructed.*

Theorem 4.2. *Let (q, t) be the generalized configuration labeling the vertex σ of the Karp and Miller tree associated with the initial configuration (p, x). For each integer N, there exists a vector y in $\mathbf{N}^m$ such that:*

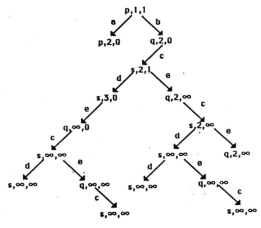

Figure 4.3

(i) $t_i = \infty \Rightarrow y_i \geq N$

(ii) $t_i < \infty \Rightarrow y_i = t_i$

(iii) *The configuration (q,y) is positively accessible from (p,x).*

Before giving the proof of theorem 4.1, we make the following observation.

(Obs. 1) *Inheritance of infinite coordinates.* If the vertex σ is an ancestor of vertex σ' in T, labeled, respectively, (p, x) and (p', x'), then each infinite coordinate of x is also an infinite coordinate of x'.

PROOF OF THEOREM 4.1.   If σ is a vertex of a Karp and Miller tree T, we denote by $(q(\sigma), P(\sigma))$ the generalized configuration labeling σ. The tree T is locally finite, since each vertex has, by construction, at most $|A|$ sons. Suppose, by way of contradiction, that T is infinite. By König's lemma (theorem 2.4), there exists an infinite branch in T. By theorem 2.5(i), we can find an infinite sequence (σ_n) of vertices in T such that, for each n, σ_n is the ancestor of σ_{n+1}, and $P(\sigma_n) \leq P(\sigma_{n+1})$. Since Q is finite, we can assume that $q(\sigma_n) = q$ does not depend on n. For no n do we have $P(\sigma_n) = P(\sigma_{n+1})$, unless σ_{n+1} has no son, by rule II.a (but does have, however, son σ_{n+2}). We then have $P(\sigma_n) < P(\sigma_{n+1})$ for all n. We show that for all n, $P(\sigma_{n+1})$ has strictly more infinite coordinates than $P(\sigma_n)$ which will give us a contradiction, since there is only a finite number of coordinates: this will conclude the proof.

Let us suppose that $P(\sigma_{n+1})$ has no more infinite coordinates than $P(\sigma_n)$. Since $P(\sigma_n) < P(\sigma_{n+1})$, they have the same infinite coordinates, and there exists an $i \in \{1, \ldots, m\}$, such that $P(\sigma_n)_i < P(\sigma_{n+1})_i < \infty$. Let σ be the father of σ_{n+1}, labeled by (q', y) and let $a \in A$ be the label of $\sigma \to \sigma_{n+1}$. By (Obs. 1), $P(\sigma_n)$, y and $P(\sigma_{n+1})$ have the same infinite coordinates. Furthermore, if, by construction, j is one of the finite coordinates of the tree $P(\sigma_{n+1})_j = (y + v(a))_j$. It follows that $y + v(a) = P(\sigma_{n+1})$. So, $P(\sigma_n) \le y + v(a)$ and $P(\sigma_n)_i < (y + v(a))_i$ from which, by rule II.b(i), we have $P(\sigma_{n+1})_i = \infty$, which is a contradiction (see figure 4.4). □

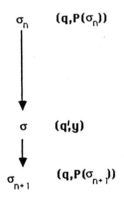

Figure 4.4

Theorem 4.2 is intuitively rather obvious. Its proof, however, is rather complicated. The principal difficulty comes from the following fact: in the tree, there may be a path whose start is the root, and which is labeled by a VASS path c, but for which c is not admissible for x (where (p, x) is the initial configuration). This fact is a result of introducing ∞ into some coordinates.

This difficulty, and is resolution, are illustrated in the following example.

Example 3

Figure 4.5 shows a VASS with one state. The Karp and Miller tree associated with the initial configuration $(0,0)$ is the one shown in figure 4.6 (we omit mention of the unique graph state in the configuration). In the tree, there is a path from the root labeled ab. However, ab is not admissible for $(0,0)$, since $(0,0)+v(ab) = (-1,1)$ which is not ≥ 0. On the other hand, the path $a^{2n}b^n$ is admissible for $(0,0)$ for any integer n: we have $(0,0)+v(a^{3n}b^n) = (n,n)$. We thus obtain the conclusion of theorem 4.2 in this particular case.

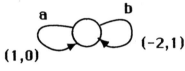

Figure 4.5

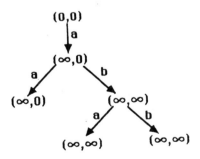

Figure 4.6

In order to prove theorem 4.2 we need a series of elementary observations and a few lemmas.

(Obs. 2) *Determinism with respect to A.* Each edge in a Karp and Miller tree T is labeled with a VASS arc, i.e., an element of A. More generally, each path in T has a label in A^* which is a VASS path obtained by the concatenation of the edge labels. Moreover, a path in

T is completely determined by its start in T and by its label: in effect, for each vertex s of T, and for each arc a in A, there is at most one arc in T labeled a which leaves s.

(Obs. 3) *From paths in a tree to paths in a VASS.* If $\gamma: \sigma \to \tau$ is a path in T labeled by a path c in a VASS, and if (p, x) and (q, y) are the labels of σ and τ, respectively, then the start of c is p and its end is q.

(Obs. 4) *Transmission of finite coordinates.* If $\gamma: \sigma \to \sigma'$ is a path in T labeled by $c \in A^*$, then if (p, x) and (p', x') are the labels of σ and σ' respectively, and if j is a finite coordinate of x', then $x'_j = x_j + v(c)_j$.

We extend the notion of admissibility of a VASS path to points in $(\mathbf{Z} \cup \infty)^m$: we will say that a VASS path c is *admissible with respect to* $I \subset \{1, \ldots, m\}$ for the point x in $(\mathbf{Z} \cup \infty)^m$ if, for every factorization $c = c_1 c_2$, we have $(x + v(c_1))_I \geq 0$.

It is clear that if c is admissible with respect to I, and if $y_I \geq x_I$, c is admissible for y with respect to I.

(Obs. 5) *From extended admissibility to admissibility.* Let c be a VASS path admissible for $x \in (\mathbf{N} \cup \infty)^m$, and let y be a vector in $\mathbf{Z}^m$ which coincides with x on the set I of the finite coordinates of x. Then c is admissible for y with respect to I.

(Obs. 6) *Concatenation of admissible paths.* Let $c_1, \ldots, c_n$ be consecutive paths in a VASS, let I be a set of coordinates, and let $x_0, \ldots, x_n \in (\mathbf{Z} \cup \infty)^m$, such that $x_i = x_{i-1} + v(c_i)$, and such that c_i is admissible for x_{i-1} with respect to I $(i = 1, \ldots, n)$. Then $c_1 \cdots c_n$ is admissible for x_0 with respect to I. Furthermore, the final point in the walk $(c_1 \cdots c_n, x_0)$ is x_n.

(Obs. 7) *Iteration of an admissible path.* Let c be a path admissible for x with respect to I, and such that $v(c)_I \geq 0$, then, for all integers n, c^n is admissible for x with respect to I.

(Obs. 8) *Obtaining admissible paths.* Let c be a VASS path, and let I be a set of coordinates. There exists an integer f such that, for each vector $z \in (\mathbf{Z} \cup \infty)^m$, $z_I \geq (f, \ldots, f)$ implies that the path c is admissible for z with respect to I. Indeed, let $x_0, \ldots, x_n$ be the intermediate points of the walk $(c, 0)$. We can then take f to be such that for all $j = 0, \ldots, n$, we have $(x_j)_I - (f, \ldots, f) \geq 0$.

Lemma 4.3. *Let* $c: q \to q$ *be a closed path in a VASS, and let* I *be a set of coordinates. There exists an integer* f *such that, for each* z

in $\mathbf{Z}^m$, and for all n in $\mathbf{N}$, $z_I \leq (f_n, \ldots, f_n)$ implies that the path c^n is admissible for z with respect to I.

PROOF. Let $f \geq 0$ be an integer such that, for all intermediate points x in the walk $(c, 0)$, and for all coordintes i in I, we have $x_i + f \geq 0$. By (Obs. 8) we then have $z_I \geq (f, \ldots, f) \Rightarrow c$ is admissible for z with respect to I.

We argue by induction on n, assuming that $z_I \geq (f_n, \ldots, f_n) \Rightarrow c^n$ is admissible for z with respect to I. Then, let z be such that $z_I \geq (f(n+1), \ldots, f(n+1))$. By the induction hypothesis, since $f \geq 0$, c^n is admissible for z with respect to I. The final point, t, of the walk (c^n, z) is $t = z + v(c^n)$, so $t_I = z_I + nv(c)_I \geq (f(n+1), \ldots, f(n+1)) + n(-f, \ldots, -f)$, since, $v(c)$ being a final point (and therefore an intermediate one) of the walk $(c, 0)$, we have $v(c)_I + (f, \ldots, f) \geq 0$. So $t_I \geq (f, \ldots, f)$, and c is admissible for t with respect to I. By (Obs. 6), we obtain that c^{n+1} is admissible for z with respect to I. □

Lemma 4.4.   *Let σ and τ be two vertices in a Karp and Miller tree, and let $c \in A^*$ be the label of the path $\sigma \rightarrow \tau$ in the tree. Let x (respectively, y) be the point in the configuration labeling σ (respectively, the father of σ). Then c is admissible for x with respect to the set of coordinates $\{j | x_j = \infty \text{ or } y_j \neq \infty\}$.*

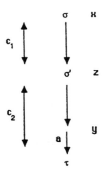

Figure 4.7

PROOF. (See figure 4.7.) If $x_j = \infty$, c is clearly admissible for x with respect to j. Let j be such that $y_j \neq \infty$. Let $c = c_1 c_2$ be a factorization of c. Assume that $c_1 \neq c$: then there exists a unique vertex σ' in the tree such that c_1 is the label of the path $\sigma \to \sigma'$, and σ' is an ancestor of the father of τ (or equal to it). Let z be the point in the configuration labeling σ'. Then by (Obs. 1) we have $z_j \neq \infty$, for $y_j \neq \infty$. So, by (Obs. 4), $z_j = x_j + v(c_1)_j$. So, $(x + v(c_1)_j)) \geq 0$.

Now, suppose that $c_1 = c$. We have $c = c'a$, where a is the last arc in c. By the previous part, we have $y_j = x_j + v(c')_j$. Furthermore, by the construction of the tree, $y_j + v(a)_j \geq 0$. It follows that $(x + v(c))_j = x_j + v(c')_j + v(a)_j = y_j + v(a)_j \geq 0$.

Therefore, c is admissible for x with respect to j, since $x_j = v(c_1)_j \geq 0$ for each left factor c_1 of c. □

In a Karp and Miller tree, there are some special vertices, namely those for which the number of infinite coordinates increases during the construction. We will call a vertex σ of the tree, labeled by (p, x) and whose father is labelled (p', x'), an *increasing vertex* when, if the set

$$\{j|\ x_j = \infty \text{ and } x'_j \neq \infty\}$$

is non-empty. This set will then be called the set of *increasing coordinates of* σ.

Example 4

Figure 4.8

In figure 4.8, the lower vertex is an increasing one, and the increasing coordinates are 2 and 3.
∎

The following lemma is technical. It contains the kernel of the proof of theorem 4.2.

Lemma 4.5. *Let σ be an increasing vertex labeled by (p, x), let σ' be its father labeled (p', x'), let $a \in A$ be the label of the edge $\sigma' \to \sigma$, and let $z = x' + v(a) \in (\mathbf{N} \cup \infty)^m$. There then exists a VASS path $c: p \to p$ which is admissible for z and such that:*
  (i) *If j is an increasing coordinate of x, then $v(x)_j \geq 1$.*
  (ii) *If j is a finite coordinate of x, then $v(c)_j = 0$.*

Such a path, corresponding to an increasing vertex σ, will be called an *increasing path* for σ.

PROOF. 1. As $z = x' + v(a)$ and $v(a) \in \mathbf{Z}^m$, x' and z have the same infinite coordinates. So, the set of increasing coordinates of σ is $\{j \mid x_j = \infty$ and $z_j \neq \infty\}$. In particular, each increasing coordinate of σ is a finite coordinate of z.
2. Let i be an increasing coordinate of σ. We will show that there exists a VASS path $c_i: p \to p$ admissible for z such that:
  (1) $v(c_i)_i \geq 1$.
  (2) If j is a finite coordinate of z, then $v(c_i)_j \geq 0$, with strict inequality only if j is an increasing coordinate of s.

Indeed, by rule II.b(i) of the construction of the tree, there exists an ancestor τ of σ labeled (p, y) such that:
  (3) $y \leq z = x' + v(a)$.
  (4) $y_i < z_i$;
  (5) If $y_j < z_j$, then j is an increasing coordinate of σ.
For c_i, we take the VASS path which labels the path $\tau \to \sigma$ in the tree. This is a path $p \to p$ in G by (Obs. 3) – cf. figure 4.9. By lemma 4.4, c_i is admissible for y with respect to every finite coordinate of x'. Thus, by 1, c_i is admissible for y with respect to the finite coordinates of z. Finally, by (3), we have $y \leq z$, so c_i is admissible for z with respect to the finite coordinates of z; but the same is clear for its infinite coordinates. Thus, c_i is admissible for z.

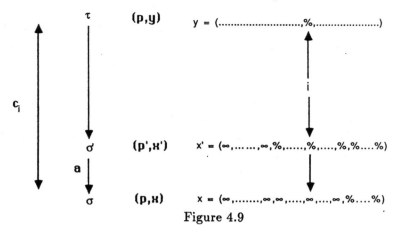

Figure 4.9

We now establish (1) and (2). We have $c_i = c_i a'$. For all finite coordinates j of z, we have $z_j = y_j + v(c_i)_j$; indeed, j is a finite coordinate of x' by 1, so by (Obs. 4), $x'_j = y_j + v(c'_i)_j$, and by definition $z_j = x'_j + v(a)_j$, from which $z_j = y_j + v(c_i)_j$. As $y_j \neq \infty$ (otherwise, by (Obs. 1), we have $x'_j = \infty$, so $z_j = \infty$ by (1)), we obtain $z_j - y_j = v(c_i)_j$, which, by (3), implies that $v(c_i)_j \geq 0$. If we have strict inequality, i.e., $v(c_i)_j > 0$, then $z_j > y_j$, so by (5), j is an increasing coordinate. This establishes (2). For (1), we note that by 1, i is a finite coordinate of z, since i is an increasing coordinate. So, the above reasoning shows that $y_i \neq \infty$, and that $z_i - y_i = v(c_i)_i$, from which (1) is derived as a consequence of (4).

3. Let $c = \prod_i c_i$, where the product is extended to all increasing coordinates. This is obviously a path, since each c_i is a path $p \to p$ in the VASS. Let j be a finite coordinate of z. As c_i is admissible for z with respect to j, and since $v(c_i) \geq 0$, we have by induction and by (Obs. 6) that c is admissible for z with respect to j. The same is clear, furthermore, if j is an infinite coordinate of z. It follows that c is admissible for z.

We have $v(c)_j = \sum_i v(c_i)_j$. By (2), this is zero if j is a finite coordinate of z which is not an increasing coordinate of σ. But these coordinates are precisely the finite coordinates of x. This proves (ii).

Now let j be an increasing coordinate of σ. Then $(v(c_j))_j \geq 1$ by (1), and $v(c_i) \geq 0$ for all other increasing coordinates i of σ. Thus $v(c)_j \geq 1$, which establishes (i). □

PROOF OF THEOREM 4.2.   1. We consider, first, the case in which s is an increasing vertex in the tree. Let $\sigma_1, \ldots, \sigma_l = \sigma$ be the increasing vertices encountered along the path from the root to σ. Let $\sigma_0 = \rho$. We argue by induction on l. Let I be the set of infinite coordinates of t, I_a the set of increasing coordinates of σ_l and let $I' = I \backslash I_a$. Futhermore, let $J = \{1, \ldots, m\} \backslash I$. Then J is the set of finite coordinates of t, and I' is the set of infinite coordinates of t, where (q', t') is the configuration which labels σ_{l-1}. See figure 4.10.

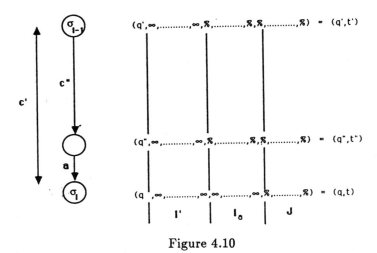

Figure 4.10

2. By induction on l, for each integer M, there exists a vector $z \in \mathbf{N}^m$, such that:

  (i') $\forall i \in I', z_i \geq M$.
  (ii') $\forall i \in I_a \cup I, z_i = t'_i$.
  (iii') (q', z) is positively accessible from (p, x).

Let c' be the VASS path labeling the path $\sigma_{l-1} \to \sigma_l$ in the tree. By lemma 4.4, c' is admissible for t' since the points in the configuration labeling σ_{l-1} and the father of σ_l have the same infinite coordinates. By (Obs.5) and (ii'), c' is admissible for z with respect to $I_a \cup I$. By (Obs. 8) and (i'), c' is admissible for z with respect to I', provided that M is large enough: in this case, c' is admissible for z.

3. Now let $c: q \to q$ be an increasing path for vertex σ_l (cf. lemma 4.5). We then have:

    (a) $v(c)_i \geq 1$, if $i \in I_a$.

    (b) $v(c)_i = 0$, if $i \in J$.

    (c) c is admissible for $t'' + v(a)$, where t'' is the point of the configuration labeling the father of σ_l, and a is the last edge of c'.

By lemma 4.3, there is an integer f such that:

    (d) For all u in $\mathbf{N}^m$, and for all n in $\mathbf{N}$, $u_{I'} \geq (fn, \ldots, fn) \Rightarrow c^n$ is admissible for u with respect to I'. We may assume that $v(c) + (f, \ldots, f) \geq 0$.

4. Now let N be an arbitrary integer, $N' \geq N$ such that $N' + v(c'_i) \geq N$ for all coordinates i, and let M be such that $M + v(c')_i \geq (f+1)N'$ for all coordinates i.

We show that $c'c^{N'}$ is admissible for z. By 2, c' is admissible for z. Furthermore, by (i'), we have $(z + v(c'))_{I'} \geq (M, \ldots, M) + v(c')_I$, so $(z + v(c'))_{I'} \geq (fN', \ldots, fN')$ by choice of M. This implies by (d) that $c^{N'}$ is admissible for $z + v(c')$ with respect to I'.

Otherwise, we write $c' = c''a$. Then, each i in $I_a \cup I$ is a finite coordinate of t'', so by (Obs. 4), $t''_i = t'_i + v(c'')_i$, from which $(t'' + v(a))_i = t'_i + v(c'')_i + v(a)_i = t'_i + v(c')_i$. But, by (c), the path c is admissible for $t'' + v(a)$, and by (ii'), $z_i = t'_i \Rightarrow (z + v(c'))_i = (t'' + v(a))_i$. So, by (Obs. 5), c is admissible for $z + v(c')$ with respect to $I_a \cup J$. Moreover, by (a) and (b), we have $v(c) \geq 0$ with respect to $I_a \cup J$, which by (Obs. 7) implies that $c^{N'}$ is admissible for $z + v(c')$ with respect to $I_a \cup J$.

Finally, $c^{N'}$ is admissible for $z + v(c')$, and c' is admissible for z. By (Obs. 6), $c'c^{N'}$ is admissible for z.

5. As (q', z) is positively accessible from (p, x) by (iii'), and $c'c^{N'}$ is admissible for z, we have, for $y = x + v(c'c^{N'})$, that (q, y) is positively accessible from (p, x).

We now verify that conditions (i), (ii) and (iii) in the statement of the theorem hold.

Let $i \in I'$: then by (i') and (d), $y_i = z_i + v(c')_i + N'v(c)_i \geq M + v(c')_i - N'f$. So, $y_i \geq N' \geq N$ by choice of N'.

Let $i \in I_a$: then by (a), $y_i \geq z_i + v(c')_i + N$, from which, by construction of N and the fact that $z \geq 0$, we obtain $y_i \geq N$.

Let $i \in J$: then $y_i = z_i + v(c')_i + N'v(c)_i = t'_i + v(c')_i$, by (ii') and (b), so $y_i = t_i$ by (Obs. 4).

6. Now assume that σ is not an increasing vertex. Let σ_l be the last increasing vertex in the path from the root to σ, and assume that c labels the path $\sigma_l \to \sigma$. Let (q', t') be the label of σ_l.

Let I be the set of infinite coordinates of $t' - t$ also has the same infinite coordinates.

Let N be an arbitrary integer. By the five previous sections of this proof, there exists a vector y' in $\mathbf{N}^m$ such that:

(i'') $\forall i \; in I, y'_i \geq N$;

(ii'') $\forall i \notin I, y'_i = t'_i$;

(iii'') (q', y') is positively accessible from (p, x).

Set $y = y' + v(c)$. For sufficiently large N, by (Obs. 7), c is admissible for y' with respect to I. But if $i \notin I$, c is admissible for t' with respect to i by lemma 4.4 (since i is finite for t, and so, also for the point which labels the father of σ, by (Obs. 1)). Since, by (ii''), $t'_i = y'_i$, c is admissible for y' with respect to i. In conclusion, c is accessible for y'. Hence, configuration (q, y) is positively accessible from (p, x) by (Obs. 6).

If $i \notin I$, we have $y_i = y'_i + v(c)_i = t'_i + v(c)_i = t_i$, by (Obs. 4). If $i \in I$, then $y_i = y'_i + v(c)_i \geq N + v(c)_i$, which completes the proof since N is arbitrary. □

4.3 Covering Graphs

The *covering graph* $\mathcal{G}$ of a VASS is obtained in the following way. We identify in the Karp and Miller tree each terminal node having an ancestor with the same label, with that ancestor.

Example 5

We take the VASS from example 2 which is reproduced in figure 4.11. Its covering graph, which can be produced directly from its Karp and Miller tree (cf. example 2), is shown in figure 4.12.

∎

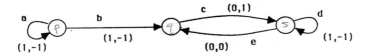

Figure 4.11

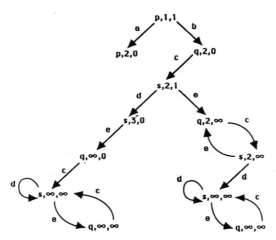

Figure 4.12

Observation (Obs. 2) of the previous section also applies to coverings: for all paths c in a VASS, and for all vertex σ in $\mathcal{G}$, there exists at most one path γ in $\mathcal{G}$ with start σ and label c.

In the case of admissible VASS paths, we have a more exact result. We denote by ρ that vertex in the covering graph which corresponds to the root of the Karp and Miller tree.

Lemma 4.6. *Let $\mathcal{G}$ be the covering graph associated with the initial configuration (p, x). Let c be a path in the VASS starting at p, and which is admissible for x. Let y be the final point in the walk (c, x). Then there exists in $\mathcal{G}$ a unique path γ with start ρ and label c. If z is the point of the configuration labeling the end of γ, then $y \leq z$.*

PROOF. (By induction on the length of c.) If c is empty, the result is obvious. If not, then c can be written $c'a$ for an arc a in A. By the induction hypothesis, there exists a path γ' in $\mathcal{G}$ whose start is ρ and whose label is c' such that the end σ' is labeled (q', z') with $y' \leq z'$, where y' is the final point of the walk (c', x). Since c is admissible for x, a is admissible for $x + v(c') = y'$, that is: $y' + v(a) \geq 0$. Thus, $z' + v(a) \geq 0$. By rule II.b of the construction of Karp and Miller trees, and by the construction of $\mathcal{G}$, there exists in $\mathcal{G}$ a successor σ of σ' such that a is the label of the edge $a: \sigma' \to \sigma$ in $\mathcal{G}$. Then $c = c'a$ is the label of the path $\gamma = \gamma'a$ in $\mathcal{G}$, so the point of the configuration labeling σ is z. By rules II.b(i) and (ii), $z \geq z' + v(a)$. Thus, $z \geq y' + v(a) = x + v(c') + v(a) = x + v(c) = y$, and the proof is complete. □

To avoid any ambiguity, we will refer to the *component* of a vector $v \in (\mathbf{Z} \cup \infty)^m$ as any one of the v_i, for some coordinate $i \in \{1, \ldots, m\}$. We will talk of *finite* and *infinite* components.

Theorem 4.7.   *Let T be a Karp and Miller tree associated with an initial configuration (p, x). We assume that T does not contain the point $(\infty, \ldots, \infty)$. Let f be the largest finite component in T. Then, for all paths c in the VASS with start p, which are admissible for x, there exists a coordinate $j \in \{1, \ldots, m\}$ such that, for each intermediate point y in the walk (c, x), we have $y_j \leq f$.*

PROOF. 1. Remark (Obs. 1) of the previous section extends to $\mathcal{G}$: if σ and σ' are two vertices of $\mathcal{G}$ labeled (q, y) and (q', y'), and if there exists in $\mathcal{G}$ a path $\sigma \to \sigma'$, then $y = \infty$ and $y'_j = \infty$.

2. The set of configurations of $\mathcal{G}$ is identical to that of T. Consequently, $\mathcal{G}$ does not contain the point $(\infty, \ldots, \infty)$ and each finite component of $\mathcal{G}$ is $\leq f$.

3. Let c be a VASS path with origin p which is admissible for x. Let $x = x_0, x_1, \ldots, x_n$ be the intermediate points of the walk (c, x). By lemma 4.6, there exists a path γ with start p and label c in γ; let $x = y_0, y_1, \ldots, y_n$ be the points of the configurations of the successive vertices in γ. Again, by lemma 4.6, we have $x_i \leq y_i$ for $i = 1, \ldots, n$. By 1 and 2, there exists $j \in \{1, \ldots, m\}$ such that $(y_i)_j \neq \infty$ for $i = 0, \ldots, n$, so $(y_i)_j \leq f$. It follows that $(x_i)_j \leq f$, which was to have been shown. □

Theorem 4.8. *Given a strongly connected VASS, a state p, a vector $x \in \mathbf{N}^m$, and a set of coordinates $J \subset \{1,\ldots,m\}$, the following property is decidable:*

(*) *There exists a vector $\Delta \in \mathbf{Z}^m$ such that $\Delta_J \geq (1,\ldots,1)$, and that $(p, x + \Delta)$ is positively accessible from (p,x) with respect to J (i.e., $(p,x) \Rightarrow_J (p, x + \Delta)$).*

Furthermore, if this property does not hold, one can effectively find an integer f such that, for all paths c of the VASS whose start is p, and which are admissible for x with respect to J, there exists $j \in J$ such that, for all intermediate points y in the walk (c, x), we have $y_j \leq f$.

PROOF. It suffices to prove the theorem for $J = \{1,\ldots,m\}$ (reduction to this case can always be made by projection onto $\mathbf{N}^J$). In this case, (*) is equivalent to the existence of the vector $(\infty,\ldots,\infty) \in \mathbf{N}^m$ in a configuration of the Karp and Miller tree associated with the initial configuration (p,x). Indeed, if such a vector exists, there exists by theorem 4.2, a state q, such that for each integer N there exists $y \geq (N,\ldots,N)$ such that $(p,x) \Rightarrow (q,y)$. But the VASS is strongly connected, so there exists a path $c:q \to p$. For sufficiently large N, c is admissible for y, and $(N,\ldots,N)+v(c) \geq (1,\ldots,1)+x$. So, we have $(p,x) \Rightarrow (p,y+v(c))$, and for $y + v(c) = x + \Delta$, we have $\Delta = y + v(c) - x \geq (N,\ldots,N) + v(c) - x \geq (1,\ldots,1)$.

Now, if $(\infty,\ldots,\infty)$ is not in the Karp and Miller tree, we use theorem 4.7: if f is the largest finite component of the points in the tree, then, for each path c starting at p which is admissible for x, there exists j such that for each intermediate point y in the walk (c,x), we have $y_j \leq f$. In particular, we could not have $(p,x) \Rightarrow (p, x + \Delta)$ for $\Delta \geq (1,\ldots,1)$, otherwise we would have for all n, $(p,x) \Rightarrow (p, x + n\Delta)$, which contradicts the previous case, since $(x + n\Delta) \geq n$ $\square$

Another consequence of theorem 4.2 and lemma 4.6 is the following classic result. We will make no use of it below, and will also leave its proof to the reader (cf. exercise 4.5).

Theorem 4.9. *Let T be the Karp and Miller tree associated with the initial configuration (p,x). The set E of configurations which are positively accessible from (p,x) is finite if and only if T does not contain the symbol ∞. In this case, E is equal to the set of configurations* *labeling the vertices of* T. *In particular, it is decidable whether* E *is finite.*

4.4 Existence of Positive Walks

The following theorem (whose statement and proof are very technical) is at the heart of Kosaraju's algorithm. A simplified version is given in exercise 4.7, which may be of use to the reader. Recall that $\gamma(c)$ is the commutative image of the path c (cf. section 2.1).

Theorem 4.10. *Let* G *be a VASS,* p *and* q *states,* Γ, Δ, x *and* y *vectors in* $\mathbf{Z}^m$, u *and* w *vectors in* $\mathbf{N}^m$, *and* g, h *integers in* $\mathbf{N}$ *such that:*

(i) *For* $i = 1, \ldots, m$, *we have* $u_i = 0$ *(respectively* $w_i = 0$*) implies* $\Gamma_i > 0$ *(respectively,* $\Delta_i > 0$*).*

(ii) *The configuration* (q, y) *is accessible from* (p, x).

(iii) $(p, x+gu+\Gamma)$ *is positively accessible from* $(p, x+gu)$ *and* $(q, y+hw)$ *is positively accessible from* $(q, y + hw + \Delta)$.

(iv) *In* G, *there is a path* c *such that* $\gamma(c) \geq (1, \ldots, 1)$ *and such that* $(p, u) \overset{c}{\to} (p, w)$.

Then there exist integers $\delta \geq 1$ *and* N *such that, for all* $k \geq N$, *where* k *is a multiple of* δ, *we have:*

$$(p, x + ku) \Rightarrow (q, y + ku)$$

If $k \mapsto M(k)$ is a mapping $\mathbf{N} \to \mathbf{N}^m$, we write $\lim M(k) = \infty$ to indicate that each component of $M(k)$ tends to infinity as k tends to infinity. The following lemma is obvious.

Lemma 4.11. *Let* $k \mapsto M(k)$ *be a mapping* $\mathbf{N} \to \mathbf{N}^m$ *such that* $\lim M(k) = \infty$. *Let* c *be a path in a VASS. Then the walk defined by* c *and the initial (respectively, final) point* $M(k)$ *is positive for sufficiently large* k.

We will also need the following result, which depends on the convexity of $\mathbf{N}^m$.

Lemma 4.12. *Let* c *be a closed VASS path and let* $x \in \mathbf{N}^m$. *For the walk* (c^n, x) *to be positive, it is sufficient for the two walks* (c, x) *and* $(c, x + (n - 1)v(c))$ *to be positive.*

PROOF. For (c^n, x) to be a positive walk, it is enough by (Obs. 6) of section 4.2 for the n walks $(c, x + iv(c))$, $0 \leq i \leq n - 1$ to be positive. But, if $x_0, \ldots, x_r$ are the intermediate points of the walk (c, x), then $x_0 + iv(c), \ldots, x_r + iv(c)$ are the intermediate points of the walk $(c, x + iv(x))$. By hypothesis, x_j and $x_j + (n - 1)v(c)$ are in $\mathbf{N}^m$. The same holds for $x_j + iv(x)$ when $0 \leq i \leq n - 1$, since this point is in the segment $[x_j, x_j + (n - 1)v(c)]$. The walk $(c, x + iv(c))$ is therefore positive. $\square$

PROOF OF THEOREM 4.10. We consider as separate the case in which q is an isolated point of the graph G.

1. Assume, then, that q is not isolated. By hypothesis, there exist in the VASS G paths c_1, c_2, c_4, such that:

$$(p, x + gu) \overset{c_1}{\Rightarrow} (p, x + gu + \Gamma)$$

$$(p, x) \overset{c_2}{\Rightarrow} (q, y)$$

$$(q, y + hw + \Delta) \overset{c_4}{\Rightarrow} (q, y + hw)$$

2. Let $\delta \geq 1$ be an integer larger than the absolute value of all the coordinates of Γ, Δ and $\gamma(c_1) + \gamma(c_4)$. We then have:
   (a) $\delta u + \Gamma \geq (1, \ldots, 1)$
   (b) $\delta w + \Delta \geq (1, \ldots, 1)$
These inequalities follow from assumption (i).

We now consider the vector $m = \delta\gamma(c) - \gamma(c_1) - \gamma(c_4) \in \mathbf{Z}^A$ (where, as usual, A denotes the set of edges of the VASS). Let a be an element of A, then $\gamma(c)_a \geq 1$ by (iv), and $\delta > (\gamma(c_1) + \gamma(c_4))_a$ by choice of δ, so $m \geq 1$, from which we derive:

   (c) $m = \delta\gamma(c) - \gamma(c_1) - \gamma(c_4) \geq (1, \ldots, 1)$.

Now, the result of m is zero because c, c_1 and c_4 are all closed paths (therefore of result zero). As $\gamma(c) \geq (1, \ldots, 1)$, each arc in G is in c. So, c passes through all the non-isolated points in G, and therefore $G \backslash \{isolated\ points\}$ is connected. Since q is not isolated, by corollary 2.2 there is a path $c_3: q \to q$ such that:

$$\gamma(c_3) = \delta\gamma(c) - \gamma(c_1) - \gamma(c_4)$$

The valuation of c_3 is then:

$$v(c_3) = \delta v(c_3) - v(c_1) - v(c_4)$$
$$= \delta(w - u) - \Gamma + \Delta$$

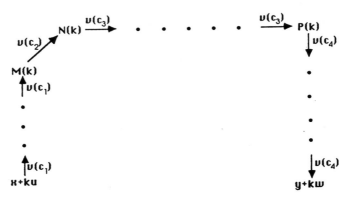

Figure 4.13

3. Consider, for $l \in \mathbf{N}$, the path

$$c(l) = c_1^l c_2 c_3^l c_4^l$$

This is indeed a path because $\alpha(c_1) = \omega(c_1) = p = \alpha(c_2)$, $\omega(c_2) = q = \alpha(c_3) = \omega(c_3) = \alpha(c_4) = \omega(c_4)$. We show that for sufficiently large k which is a multiple of δ, and for which $l = k/\delta$, the walk $(c(l), x + ku)$ is positive: the theorem will then be proved since the final point of this walk is:

$$x + ku + v(c(l)) = x + ku + lv(c_1) + v(c_2) + lv(c_3) + lv(c_4)$$
$$= x + ku + l\Gamma + y - x + l\delta(w - u) - l\Gamma + l\Delta - l\Delta$$
$$= y + kw$$

4. Let k be, then, a multiple of δ and $l = k/\delta$. Let $M(k)$, $N(k)$ and $P(k)$ be the points in $\mathbf{N}^m$ defined by:

$$M(k) = x + ku + v(c_i^l) \quad \angle \quad v(c_i)$$
$$N(k) = x + ku + v(c_1^l c_2)$$
$$P(k) = x + ku + v(c_1^l c_2 c_3^l)$$

These are particular intermediate points in the walk $(c(l), x + ku)$, see figure 4.13.

From (Obs. 6) of section 2 for $(c(l), x + ku)$ to be a positive walk, it is enough that the four walks $(c_1^l, x + ku)$, $(c_2, M(k))$, $(c_3^l, N(k))$ and $(c_4, P(k))$ be positive. By lemma 4.12, for this to be the case, it is enough that the seven walks $(c_1, x + ku)$, $(c_1, M(k) - v(c_1))$, $(c_2, M(k))$, $(c_3, N(k))$, $(c_3, P(k) - v(c_3))$, $(c_4, P(k))$ and $(c_4, y + kw - v(c_4))$ be positive.

For the first and the last walk to be positive, it is enough that $k \geq g, h$. By (a), the walks $(c_1, x + gu)$ and $(c_4, y + hw - v(c_4))$ are positive. Since $u, w \geq 0$, by assumption, the walks $(c_1, x + ku)$ and $(c_4, y + kw - v(c_4))$ are also positive if $k \geq g, h$.

For the five other walks, we apply lemma 4.11. It is enough to be shown that:

$$\lim M(k) = \infty$$
$$\lim N(k) = \infty$$
$$\lim P(k) = \infty$$

As $N(k) = M(k) + v(c_2)$, it is enough to show that the two outermost relations holds. But:

$$M(k) = x + ku + lv(c_1)$$
$$= x + l\delta u + l\Gamma$$
$$= x + l(\delta u + \Gamma)$$

and one can apply (a). In the same way:

$$P(k) = y + kw - lv(c_4)$$
$$= y + l\delta w + l\Delta$$
$$= y + l(\delta w + \Delta)$$

and (b) can be then applied.

5. We now consider the case in which y is isolated. The walk $(p, x) \to (q, y)$ of (ii) is then trivial, so $p = q$ and $x = y$. Furthermore, the path c in (iv) is trivial ($q = p$ would not be isolated otherwise), so $u = w$. The conclusion follows immediately. $\square$

Corollary 4.13.   *Let G be a VASS, p and q states, $x, y \in \mathbf{Z}^m$, $u, w \in \mathbf{N}^m$, $\Gamma, \Delta \in \mathbf{Z}^m$, $g, h \in \mathbf{N}$ and $R \subset \{1, \ldots, m\}$ such that:*

(o) *For each arc a in G, and for all j in R, the coordinate $v(a)_j$ is zero.*

(i) *For j in $\{1, \ldots, m\}\backslash R$, $u_j = 0$ (respectively, $w_j = 0$) implies $\Gamma_j > 0$ (respectively, $\Delta_j > 0$).*

(ii) *(q, y) is accessible from (p, x).*

(iii) *$(p, x + gu + \Gamma)$ is positively accessible from $(p, x + gu)$, and $(q, y + hw)$ is positively accessible from $(q, y + hw + \Delta)$.*

(iv) *There is a path c in G such that $\gamma(c) \geq (1, \ldots, 1)$ and such that $(p, u) \xrightarrow{c} (p, w)$.*

Then there exist integers $\delta \geq 1$ and N such that for all $k \geq N$, k is a multiple of δ, we have:

$$(p, x + ku) \Rightarrow (q, y + kw)$$

PROOF. By (o) and (ii), we have $x_j = y_j$ for all j in R. Furthermore, by (iv), we also have $u_j = w_j$. We consider now the VASS G' obtained by projecting the arc labels of G onto $\mathbf{N}^J$ (where $J = \{1, \ldots, m\}\backslash R$). By theorem 4.10, there exist N and δ such that, if $k \geq N$ with k a multiple of δ, we have in G':

$$(p, (x + ku)_J) \Rightarrow (q, (y + kw)_J)$$

Returning to G, we have in G by virtue of the above remarks:

$$(p, x + ku) \Rightarrow_J (q, y + kw)$$

But, if $j \in R$, and if $k \geq g$, then $(x + ku)_j = (x + gu)_j + (k - g)u_j \geq 0$ by (iii) and by the fact that $u \in \mathbf{N}^m$. Since, by (o), all the intermediate points of the walk $(p, x + ku) \rightarrow (q, y + kw)$ have the same j^{th} coordinate, we then have:

$$(p, x + ku) \Rightarrow (q, y + kw)$$

in G. $\square$

(0,0,0)

(0,1,-1)   p   q   (0,-1,2)

(1,0,0)

Figure 4.14

Exercises

4.1 Verify that, in example 1, for all vectors x in $\mathbf{N}^3$, the set of configurations which are positively accessible from the configuration $(1, x)$ is finite.

4.2 Consider the VASS shown in figure 4.14 with initial configuration (p, x) where $x = (0, 0, 1)$. Show that the set of configurations that are positively accessible from this configuration is:

$$\{(p, y)\mid 0 < y_2 + y_3 \leq 2^{y_1}\}$$
$$\cup \{(q, y)\mid 0 < 2y_2 + y_3 \leq 2^{y_1+1}\}$$

Show that the projection onto $\mathbf{N}^3$ of each of these two sets is not semi-linear (cf. Hopcroft and Pansiot, 1979, lemma 2.8)

4.3 Construct the Karp and Miller tree from the VASS in exercise 4.2.

4.4 Do the same as required by exercise 4.3 for the VASS shown in figure 4.15.

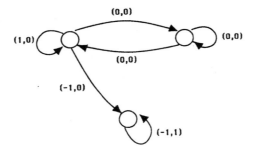

(0,0)

(1,0)   (0,0)

(0,0)

(-1,0)

(-1,1)

Figure 4.15

4.5 Prove theorem 4.9.

4.6 Construct the covering graphs for the VASSs in the preceding exercises.

4.7 A simplified version of theorem 4.10 (proved by Kosaraju, 1982, theorem 1) is as follows. Let G be a VASS labeled in $\mathbf{Z}^m$, let q_1 and q_2 be two states, and let x_1, x_2, Δ_1 and Δ_2 be vectors in $\mathbf{N}^m$ with strictly positive coordinates such that:

(i) (q_2, x_2) is accessible from (q_1, x_1).

(ii) $(q_1, x_1 + \Delta_1)$ is positively accessible from (q_1, x_1).

(iii) (q_2, x_2) is positively accessible from $(q_2, x_2 + \Delta_2)$.

(iv) $(q_2, \Delta_2 - \Delta_1)$ is accessible from $(q_2, 0)$.

Then (q_2, x_2) is positively accessible from (q_1, x_1). Hints: to go from (q_1, x_1) to (q_2, x_2), use a path of the form $c_2^j c_1 c_4^j c_3^j$, where the paths c_1, c_2, c_3 and c_4 correspond, respectively, to conditions (i), (ii), (iii) and (iv).

Notes

Theorems 4.1, 4.2 and 4.9 are by Karp and Miller (1979). Theorems 4.7 and 4.8 are hinted at by Kosaraju (1982). There exist several variants of Karp and Miller trees and covering graphs: see Jantzen and Valk (1980), Reisig (1982, 1985). We have followed Karp and Miller's original presentation here. Theorem 4.10 is by Kosaraju (1982): here, we have simplified it slightly. Kosaraju notes that one of the ideas was first presented by Sacerdote and Tenney (1977).

CHAPTER FIVE

Chains and Accessibility

In this chapter we will show that the accessibility problem is decidable. First of all, we need to introduce the concept of a chain in a VASS. Then we present the operations defined over chains: they will allow us to replace a chain by one of smaller size. In the third section, we show that a particular set (which partially describes the walks in the chain) is effectively semi-linear. This semi-linearity allows us to show that property θ is decidable (property θ is introduced in section 5.4): we will also need some results about covering graphs. Property θ implies accessibility. Conversely, in section 5.5, we show that if property θ does not hold, then the accessibility problem reduces to the accessibility problem for a finite number of chains with a length less than that of the original chain.

All of the results together constitute Kosaraju's algorithm which is given in the last section. The termination of this algorithm derives from König's lemma, and from the fact that the ordered set of chain sizes is Artin.

5.1 Definitions

Given a VASS labeled in $\mathbf{Z}^m$ and a state r, we will call a vector f in $(\mathbf{N} \cup \omega)^m$, where ω is a new symbol (whose intuitive interpretation is 'indeterminate'), a *constraint* on r. We will say that a vector z in $\mathbf{Z}^m$ *satisfies the constraint* f if, for all i in $\{1, \ldots, m\}$, we have either $f_i = \omega$ and $z_i \geq 0$, or $f_i = z_i$ (so z and f_m coincide on those coordinates of f which are $\neq \omega$). Note that z is an element of $\mathbf{N}^m$.

A *chain of vector addition systems with states* (CVASS) is a special kind of VASS with $2n$ constraints: a chain is shown schematically in figure 5.1. Each of the G_i is itself a VASS: $G_i = (Q_i, A_i)$. There are distinguished states p_i and q_i in Q_i (we can have $p_i = q_i$, it should be noted). We often denote p_1 by an arrow pointing to, and q_1 by an arrow pointing from, a chain (the *initial* and *final* states of the chain).

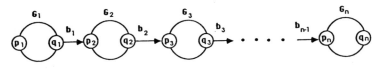

Figure 5.1

For all i, there exists a constraint e_i on p_i and a constraint s_i on q_i.

We call e_i the *entry constraint* of G_i, and we call s_i the *exit constraint* on G_i. We call p_i the *entry state* of G_i, and q_i the *exit state*. We shall write $E_i = \{j \in \{1, \ldots, m\} | (e_i)_j \neq \omega\}$ (which is the set of *coordinates constrained at entry/* of G_i), and $S_i = \{j \in \{1, \ldots, m\} | (s_i)_j \neq \omega\}$ (which is the set of *coordinates constrained at output* of G_i). We shall also write $R_i = \{j \in \{1, \ldots, m\} | \forall a \in A_i, v(a)_j = 0\}$, which we will refer to as the set of *rigid coordinates* of G_i. The size of the chain is n.

Given a path $c: p_1 \to q_n$ in a CVASS (which we denote by C), there exists one and only one factorization of c of the form:

$$c = c_1 b_1 c_2 b_2 \cdots b_{n-1} c_n$$

where each of the c_i is a path in G_i (with $\alpha(c_i) = p_i$ and $\omega(c_i) = q_i$).

Given a walk $P = (c, x)$ in G with an initial state p_1 and final state q_n, we define the *entry point* of P in G_i as:

$$x_i = x + v(c_1 b_1 \cdots b_{i-1})$$

and the *exit point* of P in G_i as:

$$y_i = x + v(c_1 b_1 \cdots b_{i-1} c_i)$$

The *intermediate points of P in G_i* are the intermediate points of the walk (c_i, x_i).

The walk P is said to be *constrained* if its initial state is p_1, its final state is q_n, and if, for all i, x_i satisfies the constraint e_i, and y_i satisfies the constraint s_i.

Example 1

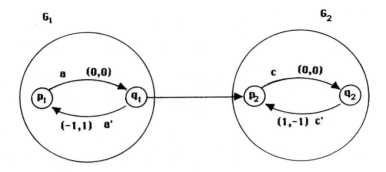

Figure 5.2

The chain C is shown in figure 5.2. The constraints here are: $e_1 = (x, y)$, $s_1 = (0, \omega)$, $e_2 = (\omega, \omega)$, and $s_2 = (\omega, 0)$.

The walk $((aa')^x ab_1 (cc')^{x+y} c, (x, y))$ has, as entry and exit points, the following:

$$x_1 = (x, y)$$

$$y_1 = (x, y) + v((aa')^x a)$$

$$= (x, y) + x(-1, 1) = (0, x + y)$$

$$x_2 = y_1 + v(b_1) = (0, x + y)$$

$$y_2 = x_2 + v((cc')^{x+y} c)$$

$$= (0, x + y) + (x + y)(1, -1) = (x + y, 0)$$

As x_1, y_1, x_2 and y_2 satisfy the constraints e_1, s_1, e_2 and s_2 respectively, the walk P is constrained. We observe that the walk is also positive.

■

The *accessibility problem* for a chain is that of knowing if there exists a constrained positive walk. The problem will be solved in this chapter, and we will see that it allows the resolution of all the accessibility problems posed so far.

With each chain, we associate its size $|C|$: it is the finite sequence: $|C| = (|G_1|, \ldots, |G_n|)$, where $|G_i|$ is the vector (a, b, c) in $\mathbf{N}^3$ defined as follows:

a is the number of non-rigid coordinates of G_i, $a = m - |R_i|$.

b is the number of arcs in G, $b = |A_i|$.

c is the number of unconstrained input coordinates to G_i + the number of unconstrained coordinates at the entry of G_i plus the number of unconstrained coordinates at the exit of G_i, $c = 2m - |E_i| - |S_i|$.

We will order $\mathbf{N}^3$ lexicographically, i.e., $(a, b, c) < (a', b', c')$ if and only if either $a < a'$, or $a = a'$ and $b < b'$, or $a = a'$, $b = b'$ and $c < c'$.

We use the ordering from section 2.3 to order the sizes. We take $A = \mathbf{N}^3$ ordered lexicographically. A letter (i.e., an element of A) is then a vector in $\mathbf{N}^3$. Each size is then a word in A^*. We define, for two words u and v in A^*, the relation:

$$u \rightarrow v$$

in the following way: $\exists x, y, w \in A^*, w \neq 1, \exists a \in A$, such that $u = xay$, $v = xwy$, and, for each letter b in w, we have $a > b$.

The relation $\xrightarrow{*}$ obtained by reflexive and transitive closure of $\rightarrow$ is, thus, an Artin ordering: this follows from theorem 2.6 and proposition 2.7: there is no infinite, strictly decreasing sequence of sizes.

Example 2

The length of the chain in example 1 is:

$$T_1 = ((2, 2, 1), (2, 2, 3))$$

If we set:

$$T_2 = ((2, 2, 1), (2, 2, 2), (2, 2, 1), (1, 6, 7))$$

we have:

$$T_1 \rightarrow T_2$$

We see indeed that T_2 is obtained from T_1 by replacing $(2,2,3)$ by the sequence $((2,2,2),(2,2,1),(1,6,7))$, each element of which is $< (2,2,3)$ under the lexicographic ordering on $\mathbf{N}^3$.

■

5.2 Operations on Chains

In this section we give two constructions on chains which will later allow us to perform induction on chain sizes. The constructions are of course effective.

Lemma 5.1.   *Given a chain C, an arc a in G_i, and an integer r, there exists a chain C' with a size less than that of C, and such that there is a bijection between the positive constrained walks in C whose underlying path uses the arc a exactly r times and the positive constrained walks in C'.*

The following example illustrates the relevance and the construction of lemma 5.1.

Example 3

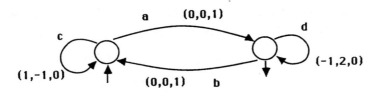

Figure 5.3

Let the chain be reduced to a single VASS (figure 5.3). The entry constraint is $e_1 = (0, 1, 0)$ and the exit constraint is $s_1 = (3, 1, 5)$. We note that $v(d)_3 = v(c)_3 = 0$, and that $v(a)_3 = v(b)_3 = 1$. This implies that, in each constrained walk (whether positive or not), each arc b is used exactly twice. We then replace the above chain by the chain of length three which simulates the constrained walks (figure 5.4—the constraints are indicated on the relevant states). The size of the first chain is $(3, 4, 0)$. The height of the new chain is $((3, 3, 3), (3, 3, 6), (3, 3, 3))$. The latter is certainly less than the former: this comes from the fact that each of the three VASSs above is obtained from the original one by removing arc b.

∎

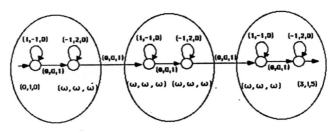

Figure 5.4

PROOF OF LEMMA 5.1.   Let p and q be the states of G_i such that $a: p \to q$. We obtain C' from C by replacing G_i by the $r + 1$ successive VASSs $G_i^1, \ldots, G_i^{r+1}$. Each of the VASSs G_i^j is obtained from G_i by suppressing the arc a. The VASS in figure 5.5 is replaced by the one in figure 5.6.

Each of the arcs a_j has the same valuation $v(a)$; the entry constraint of G_i^1 is e_i. The entry constraint of the other G_i^j is $(\omega, \ldots, \omega)$. The exit constraint of G_i^{r+1} is s_i; that of the other G_i^j is $(\omega, \ldots, \omega)$.

The G_i^j have at least as many rigid coordinates as G_i; furthermore, G_i^j has strictly fewer arcs than G_i. From this, by the lexicographic ordering, we obtain $|G_i| > |G_i^j|$, and we deduce that $|C| > |C'|$. The last assertion of the lemma is clear. ▫

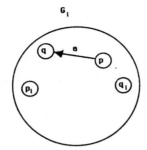

Figure 5.5

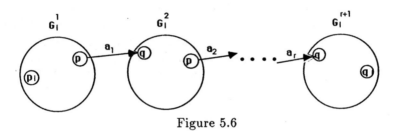

Figure 5.6

We will say that C' is obtained by *spreading C along the arc a at order r.*

Lemma 5.2. *Given a chain C, a coordinate j and an integer r, there exist chains C_{kl}, $0 \leq k, l \leq r$ such that there is a bijection between the positive constrained walks of C whose intermediate points in G_i have their j^{th} coordinate $\leq r$, and the union of the positive constrained walks in the C_{kl}. If, furthermore, j is non-rigid in G_i, then the size of each of the C_{kl} is less than that of C.*

We will say that the chains C_{kl} are obtained by *making C rigid on the coordinate j at order r.*

We illustrate lemma 5.2 with the following example.

Example 4

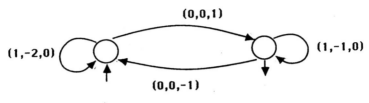

Figure 5.7

We consider the chain of order 1 shown in figure 5.7. The entry constraint is $(\omega,\omega,0)$; the exit constraint is (ω,ω,ω). For each constrained walk P, the third coordinate of P's intermediate points is equal to zero or to one. We replace the above chain, then, by the chain of length two shown in figure 5.8. The entry constraints of G' and H are, respectively, $(\omega,\omega,0)$ and (ω,ω,ω).

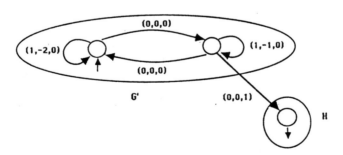

Figure 5.8

The exit constraints are (ω,ω,ω) for G' and H. The size of the first chain is $(3,4,5)$; the size of the second is $((2,4,5),(0,0,6))$ which is certainly less than $(3,4,5)$—this comes from the fact that the third coordinate has been made rigid.

■

PROOF OF LEMMA 5.2. We begin by defining a VASS G_i' whose state set is $Q_i \times \{0, 1, \ldots, r\}$, and whose arcs are defined in the following manner. There is an arc from (p, t) to (q, u) in G_i' if there is an arc a from p to q in G_i such that $u = t + v(a)_j$; the label of this arc in G_i' coincides with $v(a)$ on all coordinates except the jth which is zero.

Now, C_{kl} is obtained from C by replacing G_i by the chain of length two shown in figure 5.9. The VASS H is reduced to a single state (see figure 5.9).

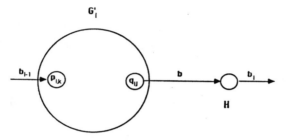

Figure 5.9

The entry constraint on G_i' is e_i; that of H is $(\omega, \ldots, \omega)$. The exit constraint of G_i is $(\omega, \ldots, \omega)$, and that of H is s_i. The arc b is labeled by the vector:

$$v(b) = (0, \ldots, 0, l - k, 0, \ldots, 0)$$

The assertion about walks can easily be verified.

Assume, now, that j is not rigid in G_i. As j is rigid in G_i', we have $|G_i| > |G_i'|$. Furthermore, all coordinates are rigid in H, so $|G_i| > |H|$, since G_i has at least one non-rigid coordinate. It follows that $|C| > |C_{kl}|$.

□

5.3 A Semi-linear Set

Let C be a chain composed of the VASSs $G_1, \ldots, G_n$. Recall that, with each constrained walk $P = (c, x)$, we associate the points x_i, y_i (the entry and exit points of P in G_i, respectively). Furthermore, c factors in a unique fashion into:

$$c = c_1 b_1 c_2 b_2 \cdots b_{n-1} c_n$$

where c_i is that part of c which is in G_i. We call the following vector:

$$(x_1, y_1, x_2, y_2, \ldots, x_n, y_n, \gamma(c_1), \ldots, \gamma(c_n))$$

$$\in \mathbf{N}^K = \mathbf{N}^{m.2n} \times \mathbf{N}^{A_1} \times \ldots \times \mathbf{N}^{A_n}$$

the *extended (commutative) image of the walk* P, where, recall γ denotes the commutative image of a path (cf. 2.1). We write ε_i for the projection of $\mathbf{N}^K$ onto $\mathbf{N}^m$ corresponding to x_i (ε_i gives, then, the entry point of P in G_i), σ_i for the projection of $\mathbf{N}^K$ onto $\mathbf{N}^m$ corresponding to y_i, and π_i for the projection of $\mathbf{N}^K$ onto $\mathbf{N}_i^A$ (π_i gives the commutative image of c_i).

We associate with the chain C the subset of $\mathbf{N}^K$ defined by:

$$L(C) = \{extended\ images\ of\ constrained\ walks\ in\ C\}$$

We remark that we include all constrained walks—not just the positive ones—in the definition of $L(C)$. The following result allows us, amongst other things, to decide whether there exists a constrained walk (which need not necessarily be positive—cf. exercise 5.4).

Theorem 5.3. $L(C)$ *is a rational (therefore semi-linear) subset of* $\mathbf{N}^K$.

This result is effectively computable, as we shall see. Recall that, in $\mathbf{N}^K$, a subset is rational if and only if it is semi-linear (proposition 3.5).

PROOF. By theorem 3.3, the set of paths through a graph with a given start and end is a rational subset of the free monoid A^* generated by the set A of arcs in the graph when each path is considered as a word over A. Consequently, the set L_1 of paths $c : p_1 \to q_n$ in the chain C is a rational subset of A^*.

Recall that A is the union of the $A_1, \ldots, A_n$ and the set $\{b_1, \ldots, b_{n-1}\}$. Let π denote the canonical projection of A^* onto B^* for $B \subset A$: i.e.,

$\pi_B|B^* = \text{id}$, $\pi_B(A^*\backslash B^*) = 1$ (the empty word). Define $\phi_i, \psi_i : A^* \to \mathbf{Z}^m$ by the following equations:

$$\phi_i = v \circ \pi_{A_1 \cup \ldots \cup A_i \cup \{b_1, \ldots, b_{i-1}\}}$$

$$\psi_i = v \circ \pi_{A_1 \cup \ldots \cup A_i \cup \{b_1, \ldots, b_i\}}$$

We define $\rho : A^* \to \mathbf{N}_i^A$ by:

$$\rho_i = \gamma \circ \pi_{A_i}$$

Then the subset $L_2 = (\phi_1, \psi_1, \ldots, \phi_n \psi_n, \rho_1, \ldots, \rho_n)(L_1)$ of $\mathbf{Z}^K$ is rational since it is the image of a rational subset under a homomorphism (cf. proposition 3.1): it consists of all the $(x_1, y_1, \ldots, x_n, y_n, \gamma(c_1), \ldots, \gamma(c_n))$, for all walks $P = (c, 0)$ in C with initial state p_1 and final state q_1, where x_i (respectively, y_i) is the entry (respectively, exit) point of P in G_i (in particular, $x_1 = 0$), and where c_i is that part of the path c contained in G_i.

We now add to L_1 the (rational) subset

$$\underbrace{\{x, x, \ldots, x, x}_{2n \text{ times}}, 0, \ldots, 0 | \; x \in \mathbf{N}^m\}$$

of $\mathbf{Z}^K$ to obtain a rational subset L_3 of $\mathbf{Z}^K$ consisting of all the

$$(x_1, y_1, \ldots, x_n, y_n, \gamma(c_1), \ldots, \gamma(c_n))$$

for all walks (c, x), using the same notation as above (but this time, $x = x_1$ is arbitrary). It remains to restrict to constrained walks. For this, let:

$$\overline{E}_i = \{x \in \mathbf{N}^m | \forall j \in \{1, \ldots, m\}, (e_i)_j \neq \omega \Rightarrow x_j = (e_i)_j\}, \quad \text{and}$$

$$\overline{S}_i = \{y \in \mathbf{N}^m | \forall j \in \{1, \ldots, m\}, (s_i)_j \neq \omega \Rightarrow y_j = (s_i)_j\}$$

In other words, $\overline{E}_i$ is the set of vectors in $\mathbf{N}^m$ which satisfy the constraints e_i. The sets $\overline{E}_i$ and $\overline{S}_i$ are clearly rational. The same is true for:

$$\overline{E}_1 \times \overline{S}_2 \times \cdots \times \overline{E}_n \times \overline{S}_n \times \mathbf{N} \cdots \times \mathbf{N} \subset \mathbf{N}^K$$

Finally, the intersection of this with L_3 is rational in $\mathbf{N}^K$ (cf. theorem 3.7 and proposition 3.6), and is precisely the set $L(C)$. □

5.4 The Property Θ

In this section we define a property of chains which ensures the existence of positive constrained walks.

Let C be a chain of length n composed of the VASSs $G_1, \ldots, G_n$ as specified in section 5.1.

We will say that C has the $\theta 1$ property if, for each integer N, there exists a constrained walk $P = (c, x)$ such that:

(i) For each arc a in $A_1 \cup \ldots \cup A_n$, the path c contains the arc a at least N times.

(ii) For all $i \in \{1, \ldots, n\}$, each unconstrained coordinate of the entry point of P in G_i is $\geq N$, and each unconstrained coordinate of the exit point of P in G_i is $\geq N$.

We will say that a chain C has the $\theta 2$ property if, defining e_i' (respectively, s_i') $\in \mathbf{N}^m$ as the vector that coincides with e_i (respectively, s_i) on all its coordinates not equal to ω, and equal to 0 on all other coordinates, we have:

(i) For all $i \in \{1, \ldots, n\}$, there exists Γ_i in $\mathbf{Z}^m$, such that if $j \in E_i \backslash R_i$ then $(\Gamma_i)_j \geq 1$, and that $(p_i, e_i') \Rightarrow_{(E_i \backslash R_i)} (p_i, e_i' + \Gamma_i)$ in the VASS G_i.

(ii) For all $i \in \{1, \ldots, n\}$, there exists Δ_i in $\mathbf{Z}^m$, such that if $j \in S_i \backslash R_i$ then $(\Delta_i)_j \geq 1$, and that $(q_i, s_i') + \Delta_i \Rightarrow_{(S_i \backslash R_i)} (q_i, s_i')$ in the VASS G_i.

Finally, we say that the chain C *has the property* θ if it has both $\theta 1$ and $\theta 2$ properties. This property is decidable by virtue of the following theorem:

Theorem 5.4.   *We can decide whether a given chain has the property* θ .

PROOF. We have to show that we can separately test for properties $\theta 1$ and $\theta 2$. Property $\theta 1$ is equivalent to the following (using notations from section 5.3): for each integer N, there exists an x in $L(C)$ such that $\forall i \in \{1, \ldots, n\}$, we have $\pi_i(x) \geq (N, \ldots, N)$, $(\varepsilon_i(x))_{\{1, \ldots, m\} \backslash E_i} \geq (N, \ldots, N)$, and $(\sigma_i(x))_{\{1, \ldots, m\} \backslash S_i} \geq (N, \ldots, N)$. It is sufficient, then, to apply proposition 3.8.

The decidability of property $\theta 2$(i) follows immediately from theorem 4.8, once G_i is replaced by the strongly connected component of p_i in G_i.

Similarly for $\theta 2$(ii), by replacing G_i by the dual VASS obtained by replacing each arc $a: p \to q$ labeled $v(a)$ in G_i by the arc $a: q \to p$ labeled $-v(a)$. □

Interest in the property θ is justified by the following result.

Theorem 5.5. *If the chain C has the property θ, then there exists a positive constrained walk.*

The proof of this theorem is very technical and uses many of the results given in this book.

PROOF. 1. As C has the property $\theta 1$, by proposition 3.8, and by the definition of $L(C)$, there exist $w, t \in \mathbf{N}^K$ such that $w + Nt \subset L(C)$ and $\forall i \in \{1, \ldots, n\}$: $\pi_i(t) \geq (1, \ldots, 1)$, $\varepsilon_i(t)_j \geq 1$ if and only if $j \in \{1, \ldots, m\} \backslash E_i$, and $\sigma_i(t)_j \geq 1$ if and only if $j \in \{1, \ldots, m\} \backslash S_i$.

Let $L_i = (\varepsilon_i, \sigma_i, \pi_i)(L(C)) \subset \mathbf{N}^{2m} \times \mathbf{N}^{A_i}$ (i.e., L_i gives the entry and exit points in G_i and the commutative images of the parts in G_i of the constrained walks in G_i). So L_i contains $(x_i, y_i, \eta_i) + N(u_i, w_i, M_i)$, where $x_i = \varepsilon_i(w)$, $y_i = \sigma_i(w)$, $\eta_i = \pi_i(w)$, $u_i = \varepsilon_i(t)$, $w_i = \sigma_i(t)$, $M_i = \pi_i(t)$. By the above remarks, we have:

$$(*) \begin{cases} (u_i)_j \geq 1 & \text{if } j \notin E_i \\ (u_i)_j = 0 & \text{if } j \in E_i \\ (w_i)_j \geq 1 & \text{if } j \notin S_i \\ (w_i)_j = 0 & \text{if } j \in S_i \\ M_i \geq (1, \ldots, 1) \end{cases}$$

2. We show that G_i satisfies the assumptions of corollary 4.3 (where p is replaced by p_i, q by q_i, and so on):

(o) For all $j \in R_i$ and all arcs $a \in A_i$, we have $v(a)_j = 0$ by definition of R_i.

(i) Let $j \notin R_i$ such that $(u_i)_j = 0$: then, by $(*)$, $j \in E_i$, so by $\theta 2$(i), $(\Gamma_i)_j \geq 1$. In the same way, if $j \notin R_i$ and $(w_i)_j = 0$, then $(\Delta_i)_j \geq 1$.

(ii) By definition of L_i, for all natural numbers r, there exists a path $c(r)$ in G_i with commutative image $\gamma(c(r)) = \eta_i + rM_i$, and such that $(p_i, x_i + ru_i) \overset{c(r)}{\to} (q_i, y_i + rw_i)$, and such that $x_i + ru_i$ satisfies the constraints e_i, and $y_i + rv_i$ satisfies constraint s_i. In particular, for $r = 0$, we obtain that (q_i, y_i) is accessible from (p_i, x_i).

(iii) By $\theta 2$(i), there exists a path c' in G_i such that $(p_i, e'_i) \overset{c'}{\Rightarrow}_{(E_i \backslash R_i)}$ $(p_i, e'_i + \Gamma_i)$. As $e'_i \geq 0$, and the coordinates in R_i of the valuations of the arcs in G_i are zero, we have $(p_i, e'_i) \overset{c'}{\Rightarrow}_{E_i} (p_i, e'_i + \Gamma_i)$. But, if $j \in E_i$, $(e'_i)_j = (e_i)_j = (x_i)_j$ (since x_i satisfies constraint e_i): thus, $(p_i, x_i) \overset{c'}{\Rightarrow}_{E_i} (p_i, x_i + \Gamma_i)$. Now, if $j \notin E_i$, we have, by $(*)$, $(u_i)_j \geq 1$; furthermore, $u_i \geq 0$. So, by (Obs. 8) of section 4.2, we have that, for sufficiently large g, $(p_i, x_i + gu_i + \Gamma_i)$ is positively accessible for $(p_i, x_i + gu_i)$.

The other part of (iii) is proved symmetrically, replacing G_i by its dual VASS.

(iv) By $(*)$, we have $M_i \geq (1, \ldots, 1)$. We show that there exists a path c in G_i such that $\gamma(c) = M_i$, and such that $(p_i, u_i) \overset{c}{\rightarrow} (p_i, w_i)$. Firstly, (cf. (ii)), there exists in G_i the path $c(1): p_i \rightarrow q_i$ with commutative image $\eta_i + M_i$, which thus has all the arcs of G_i. This shows that the graph $G'_i = G_i \backslash \{$ *isolated points*$\}$ is connected and contains p_i and q_i.

Since $M_i \geq (1, \ldots, 1)$, we have $G_i | M_i = G'_i$ (see section 2.1 for the definition of $G|M_i$). Furthermore, the value (*ibid.*) of M_i is zero, since it is the difference of the equal results of to η_i and $\eta_i + M_i$ (since η_i and $\eta_i + M_i$ are the commutative images of the paths $c(0)$, $c(1): p_i \rightarrow q_i$, cf. (ii)).

Then, by corollary 2.2, there exists a path $c: p_i \rightarrow p_i$ such that $\gamma(c) = M_i$. As $\gamma(c) = \gamma(c(1)) - \gamma(c(0))$, we have $v(c) = v(c(1)) - v(c(0)) = (y_i + w_i - x_i - u_i) - (y_i - x_i) = w_i - u_i$, from which we obtain that $(p_i, u_i) \overset{c}{\rightarrow} (p_i, w_i)$. We remark that the existence of c implies that $G \backslash \{$ *isolated points*$\}$ is strongly connected, and that we have only made use of property $\theta 1$ (and not $\theta 2$) to establish this fact.

3. By corollary 4.13, there exists, for all $i \in \{1, \ldots, n\}$, an integer $\delta_i \geq 1$ and an integer N_i such that if $k \geq N_i$, with k a multiple of δ_i, then there exists a path c_i such that:

$$(p_i, x_i + ku_i) \overset{c_i}{\Rightarrow} (q_i, y_i + kw_i) \tag{1}$$

Now, let k be a common multiple of the δ_i, and let it be larger than any of the $N_1, \ldots, N_n$. We then have (1) for all $i \in \{1, \ldots, n\}$. On the other hand, by the definition of $L(C)$, there is a walk $P = (c, x)$ whose extended image is $w + kt$, that is:

$$(x_1 + ku_1, y_1 + kw_1, \ldots, x_n + ku_n, y_n + kw_n, \eta_1 + kM_1, \ldots, \eta_n + kM_n)$$

In particular, for $i = 1, \ldots, n-1$, we have $x_{i+1} + ku_{i+1} = y_i + kw_i + v(b_i)$. This shows that the entry and exit points of the walk $(c_1 b_1 c_2 \ldots b_{n-1} c_n, x_1 + ku_1)$ are $x_i + ku_i$ and $y_i + kw_i$. By observation (Obs. 6) of section 4.2, we have the conclusion that it is a positive walk. Furthermore, since $x_i + ku_i$ and $y_i + kw_i$ satisfy the constraints, it is also a constrained walk. □

5.5 Size Reduction

In this section, we show that if the property θ is not satisfied by a chain C, then we can reduce the length of the accessibility problem.

Theorem 5.6. *Let C be a chain which does not have the property θ. Then there exists an infinite set $\mathcal{C}$ such that:*
(i) The length of each $C' \in \mathcal{C}$ is strictly less than that of C.
(ii) There exists a constrained positive walk in C if and only if there exists a positive constrained walk in one of the $C' \in \mathcal{C}$.

The constructions are effective.

PROOF. 1. If $L(C) = \emptyset$, i.e., if there are no constrained walks, we merely take $\mathcal{C} = \emptyset$. For the remainder of the proof, assume $L(C) \neq \emptyset$. There are two cases to be considered:
- C does not have the property $\theta 1$.
- C has property $\theta 1$, but not $\theta 2$.

2. C does not have property $\theta 1$. Then there exists an integer N such that, for all constrained walks $P = (c, x_0)$, we have one of the following three cases:
I. There exist $i \in \{1, \ldots, n\}$ and $a \in A_i$ such that c contains the arc a at most N times.
II. There exist $i \in \{1, \ldots, n\}$ and coordinate $j \notin E_i$ such that the entry point x_i of P in G_i satisfies $(x_i)_j \leq N$.
III. There exist $i \in \{1, \ldots, n\}$ and coordinate $j \notin S_i$ such that the exit point of P in G_i satisfies $(y_i)_j \leq N$.

We note that N is effectively computable: writing $L(C) = \bigcup_h (w_h + V_h^*)$ (finite union), we have, by hypothesis, for all h, that there exists a coordinate $j \in \{1, \ldots, K\}$ corresponding either to an arc or to an unconstrained coordinate such that $(\sum_{t \in V_h} t)_j = 0$ (cf. proposition 3.8). We can then set N to be the maximum of the $(w_h)_j$.

For $a \in \bigcup A_i$ and $k \in \{0, \ldots, N\}$, we define the chain $C_{a,k}^I$ as the chain obtained by spreading C along the arc a at order k. The size of $C_{a,k}^I$ is less than that of C (lemma 5.1).

Now let $i \in \{1, \ldots, n\}$, $j \in \{1, \ldots, m\} \backslash E_i$ and let $k \in \{0, \ldots, N\}$. The chain $C_{i,j,k}^{II}$ is defined as being identical to C except that the entry constraint vector e_i is replaced by $\overline{e}_i$ defined by:

$$(\overline{e}_i)_h = \begin{cases} k, & \text{if } h = j \\ (e_i)_h, & \text{if } h \neq j \end{cases}$$

It is clear that the length of $C_{i,j,k}^{II}$ is strictly less than that of C.

Let $i \in \{1, \ldots, n\}$, $j \in \{1, \ldots, m\} \backslash S_i$, and let $k \in \{0, \ldots, N\}$. The chain $C_{i,j,k}^{III}$ is defined symmetrically by:

$$(\overline{s}_i)_h = \begin{cases} k, & \text{if } h = j \\ (s_i)_h, & \text{if } h \neq j \end{cases}$$

In the same way as above, we have $|C| > |C_{i,j,k}^{II}|$. By lemma 5.1, we see, then, that condition (ii) of the theorem is satisfied when we take the set of chains defined above as the finite set $\mathcal{C}$ of chains.

3. C has property $\theta 1$, but not $\theta 2$.

As C has property $\theta 1$, each of the graphs $G_i' = G_i \backslash \{isolated\ points\}$ is strongly connected (cf. the remark at the end of part two of the proof of theorem 5.5; see also exercise 5.1). As C does not satisfy property $\theta 2$, C does not satisfy property $\theta 2(\text{i})$, or does not satisfy property $\theta 2(\text{ii})$. We consider the first case: the other follows by symmetry (replace C by the reverse chain). By theorem 4.8, then, there exists an integer N which is effectively computable and there exists an $i \in \{1, \ldots, n\}$ such that for each path c in G_i with start p_i which is admissible for e_i' with respect to $E_i \backslash R_i$, there exists a coordinate j in $E_i \backslash R_i$ such that, for each intermediate point z in the walk (c, e_i'), we have $z_j \leq N$. Then let $\mathcal{C}$ be the set of chains obtained by making C rigid on the different coordinates $j \in E_i \backslash R_i$ at order N.

Each one of these chains has a size less than that of C (lemma 5.2). Furthermore, if $P = (c, x_0)$ is a positive constrained walk, c_i the part of c in G_i, and if x_i is the entry point of P in G_i, then x_i and e_i' coincide on the coordinates in $E_i \backslash R_i$. Therefore, the intermediate points of the walks (c_i, x_i) and (c_i, e_i') have the same projection onto $\mathbf{N}^{E_i \backslash R_i}$, and it follows that c_i is admissible for e_i' with respect to $E_i \backslash R_i$. It follows that there exists $j \in E_i \backslash R_i$ such that the intermediate points of P in G_i have their j^{th} coordinate $\leq N$. We thus obtain, then, condition (ii) of the theorem by an application of lemma 5.2. □

5.6 The Decidability of Accessibility

The problem of accessibility in Petri nets reduces to the problem of accessibility in a vector addition system with states, as we saw in Chapter One. The accessibility problem in a VASS reduces to the problem of deciding whether there exists a positive constrained walk in a VASS chain. Indeed, let G be a VASS, p and q two states, and x and y two vectors in $\mathbf{N}^m$. Let C be the chain of length 1 reduced to the single VASS G, and let p be the initial and q be the final state; let the entry constraint be x and the exit constraint be y. It is then clear that we have:

$$(p, x) \Rightarrow (q, y)$$

in G if and only if there exists a positive constrained walk in C.

We now show how one can decide about the existence of a positive constrained walk in a given chain C. We begin by testing if C has the property θ: this is decidable (theorem 5.4). If C has the property θ, then there exists a positive constrained walk, as theorem 5.5 tells us. If C does not have the property θ, we create a *finite number* of chains with size *less* than that of C such that C admits a positive constrained walk if and only if one of the created chains has a positive constrained walk (theorem 5.6). We perform this sequence of operations on each of the newly created chains. This algorithm terminates. If it did not, by König's lemma (theorem 2.4), there would exist an infinite, strictly decreasing sequence of sizes – but this would contradict the Artin property of the ordered set of chain sizes (theorem 2.6 and proposition 2.7).

Exercises

5.1 Show that if a chain has no constraints (i.e., all the constraint vectors are equal to $(\omega, \ldots, \omega)$), the accessibility problem is equivalent to the existence problem for paths from the initial to the final state.

5.2 Describe the chains whose size contains only 0s. To what does the accessibility problem reduce in this case, and to what does the property θ reduce?

5.3 Show that for every semi-linear subset L of $\mathbf{N}^m$, there exists a VASS G and states p and q such that $L = \{x \in \mathbf{N}^m \,|\, (p, 0) \Rightarrow (q, x)\}$.

5.4 Using theorem 5.3, show that the existence of a constrained walk in a chain is decidable. This result also gives a necessary condition of accessibility in VASSs and Petri nets.

5.5 Let C be a chain for which one VASS, G_i, say, is not strongly connected. Show that the accessibility problem, in this case, can be reduced to one which deals only with a finite number of chains whose size is less than that of C.

Notes

The theorem on the decidability of accessibility in Petri nets and VASSs is due to Mayr (1981, 1984) and to Kosaraju (1982). As the reader will already have observed, the result has a double complexity – in the proof and in the algorithm. One should, rather, use the term 'non-undecidability', as does Straubing (1983, p. 281) when talking about the decidability of Burnside's problem for matrix semigroups. All of the intermediate results in this chapter are from Kosaraju (1982). Partial solutions of the accessibility problem were obtained by various authors. In particular, the accessibility set is effectively semi-linear in certain cases: see van Leeuwen (1974), Landweber and Robertson (1978), Hopcroft and Pansiot (1979), Grabowski (1980), Müller (1980, 1981). For other references to the accessibility problem, the reader may consult Müller (1982, 1985).

CHAPTER SIX

Complements

In this chapter we present some additional results. Firstly, VASSs (and Petri nets) whose associated language is recognizable are considered: this amounts, essentially, to an interation condition. Secondly, we show that the problem of the equivalence of accessibility sets for Petri nets is undecidable: this reduces to the undecidability of Hilbert's tenth problem. Finally, we prove the decidability problem for liveness of transitions.

6.1 The Language Associated with a Vector Addition System

Let $G = (Q, A)$ be a vector addition system with states which has a valuation function $v: A \to N^m$, and let $(p, x) \in Q \times N^m$ be an initial configuration.

Recall that each path in G is represented by a word in the free monoid A^* generated by A. Furthermore, a language is a subset of A^*. The *language associated* with the VASS is the set of paths $c \in A^*$ with start p and which are admissible for x.

Example 1

The language associated with the VASS shown in figure 6.1, which has one state p and whose initial configuration is $(p, 0)$ is the set of words w over the alphabet $\{a, b\}$ in which all left factors contain at least as many as as bs.

∎

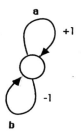

Figure 6.1

Below, we assume that the reader has some familiarity with finite state automata and recognizable languges. The aim of this section is to characterize those VASSs whose language is recognizable.

Let $G = (S, A)$ be a VASS, and let (p, x) be an initial configuration. In the associated Karp and Miller tree, T, let ϕ be a leaf labeled (q, y). If ϕ possesses an ancestor σ also labeled (q, y), let $c \in A^*$ be the label of the path $\sigma \to \phi$. We will say that T *has the (VVN) property* if we have:

$$v(c) \geq 0 \qquad\qquad (VVN)$$

and that this holds for every leaf ϕ in the tree (we write, as usual, 0 in place of $(0, \ldots, 0)$).

Example 2

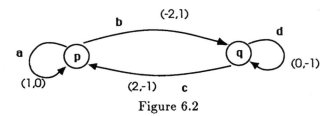

Figure 6.2

The Karp and Miller tree associated with the VASS shown in figure 6.2 is depicted in figure 6.3. The initial configuration is $(p, 0, 0)$. The

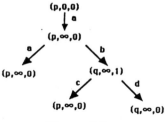

Figure 6.3

VVN condition is satisfied because we have $v(a) = (1,0) \geq 0$ and $v(bc) = (0,0) \geq 0$.

■

Theorem 6.1. *The language associated with a VASS and with an initial configuration is recognizable if and only if its Karp and Miller tree has the (VVN) property.*

This result shows, in particular, that one can decide whether the associated language is recognizable. The proof will show that one can also effectively construct a finite automaton for the language.

In terms of rational expressions, the language of the VASS in example 2 with initial configuration $(p, 0, 0)$ is:

$$aa\{a, bc\}^*\{1, bd\}$$

By Kleene's theorem, this language is recognizable.

Before giving the proof of theorem 6.1, we state and prove a lemma.

Lemma 6.2. *Let G be a VASS, and let (p, x) be an initial configuration such that the (VVN) condition is satisfied. Then there exists a computable natural number N such that for all paths c with state p and which are admissible for x, and for all factorizations $c = c_1 c_2$, we have $v(c_2) \geq -(N, \ldots, N)$.*

PROOF. Let $\mathcal{G}$ be the covering graph of G with initial configuration (p, x) (cf. 4.3). As T has the (VVN) property, for each elementary circuit of $\mathcal{G}$ (i.e., for each closed path without repeated vertices) labeled by $c \in A^*$, one has $v(c) \geq 0$. Let n be the number of vertices of $\mathcal{G}$ and let

u be the function which associates $u(\gamma) = v(c)$ with each path γ in $\mathcal{G}$ labeled by $c \in A^*$.

Then, for each path γ in $\mathcal{G}$, there exists a path γ' with length at least n such that $u(\gamma) \geq u(\gamma')$. Indeed, if $|\gamma| \geq n$, then γ factors into $\gamma = \gamma_1 \gamma' \gamma_2$, where γ' is an elementary circuit. Then $u(\gamma) = u(\gamma_i \gamma_2) + u(\gamma') \geq u(\gamma_1 \gamma_2)$, since $u(\gamma') \geq 0$ by the above, and we complete the proof by induction on $|\gamma|$.

We infer by induction that, for each path c of G with start p, and which is admissible for x, and for all factorizations $c = c_1 c_2$, we have $v(c_2) \geq v(c')$ for a path c' in G of length at most n. Indeed, by lemma 4.6, there exists a path γ in $\mathcal{G}$ with label c; the factorization $c = c_1 c_2$ corresponds to a factorization $\gamma = \gamma_1 \gamma_2$, where c_i is the label of γ_i. In particular, $u(\gamma_i) = v(c_i)$. By the above, there exists a path γ' such that $|\gamma'| \leq n$ and $u(\gamma_2) \geq u(\gamma')$. Let c' be the label of γ', then let $|c'| \leq n$, and $v(c_2) \geq u(\gamma') = v(c')$.

To finish the proof, we take N to be an integer greater than the absolute values of the negative coordinates of the $v(c')$, where c' is a path in G with start p and with length $\leq n$. □

PROOF OF THEOREM 6.1.   Let $G = (Q, A)$ be a VASS, let (p, x) be the initial configuration, L the associated language, and let T be the associated Karp and Miller tree.

1. We assume that L is recognizable. The language L is recognized by an n-state finite state automaton with n states. Let ϕ, σ, (q, y) and c be as in the statement of the (VVN) property.

Let $I = i \in \{1, \ldots, m\} | y_i = \infty\}$. By lemma 4.4, c is admissible for y, and so, too, is the path c^n (which is indeed a path, for c goes from q to q), since $v(c)_i = 0$ for each coordinate $i \notin I$ (cf. (Obs. 4) of section 4.2).

For each integer N, by theorem 4.2, there exists a vector $z(N) \in \mathbf{N}^m$ such that:
(i) $\forall i \in I, z(N)_i \geq N$.
(ii) $\forall i \notin I, z(N)_i = y_i$.
(iii) (q, z) is positively accessible from (p, x).

We choose N so that c^n is admissible for $z = z(N)$ with respect to I. Then c^n is admissible for z, since it is admissible for y, and, by (ii), y and z coincide on $\{1, \ldots, m\} \backslash I$.

By (iii), there exists a path $c_1: p \to q$ which is admissible for x, and such that $(p, x) \overset{c_1}{\to} (q, z)$. Then $c_1 c^n$ is in L. By the iteration lemma for recognizable languages, since L is recognized by an n-state finite automaton, there is an integer i, $1 \le i \le n$, such that $\forall k \in \mathbf{N}, c_1 c^{n-i}(c^i)^k \in L$.

This implies that all these paths are admissible for x, so, in particular, we have:

$$\forall k \in \mathbf{N}, 0 \le x + v(c_1 c^{n-1} c^{ik}) = x + v(c_1 c^{n-1}) + ikv(c)$$

It follows that we must have $v(c) \ge 0$. This shows that L has the (VVN) property.

2. We now assume that the Karp and Miller tree of G has the (VVN) property. Let N be the integer of lemma 6.2. We may assume that $x \le (N, \dots, N)$.

We construct a deterministic finite state automaton $\mathcal{A}$ whose state set is:

$$E = \{(q, y) \in Q \times \mathbf{N}^m \mid y \le (N, \dots, N)\}$$

The initial state is (p, x). All the states are terminal. The transition function $\delta: E \times A \to E$ is defined in the following manner. We set $\delta(q, y, a) = \emptyset$ except when $\alpha(a) = q$ and $y + v(a) \ge 0$. In this case, $\delta(q, y, a) = (r, z)$, with $r = \omega(a)$ (so a is an edge of G which goes from q to r), and $z_i = \min\{y_i + v(a)_i, N\}$ for all coordinates i. This may also be written $z = \inf\{y + v(a)_i, (N, \dots, N)\}$, with the natural ordering over $\mathbf{N}^m$.

We can make three observations about this automaton:

(1) If c is in A^*, and $\delta(p, x, c) = (q, y)$, then c is a path with start p and end q in G, and y is less than or equal to the final point of the walk (c, x).

(2) If c in A^* is such that $\delta(p, x, c) = \emptyset$, and if i is a coordinate such that $(x + v(c))_i > N$, but that $(x + v(c'))_i \le N$, for all proper left factors c', then $\delta(p, x, c) = (r, z)$, with $z_i = N$.

(3) If $\delta(p_0, x_0, a_1) = (p_1, x_1)$, $\delta(p_1, x_1, a_2) = (p_2, x_2)$, ..., $\delta(p_{n-1}, x_{n-1}, a_n) = (p_n, x_n)$, with $a_1, \dots, a_n \in A$, and if, for some coordinate i, we have $(x_1)_i, (x_2)_i, \dots, (x_n)_i < N$, then $(x_n)_i = (x_0)_i + v(a_1, \dots, a_n)_i$.

Let $c \in A^*$ be recognized by $\mathcal{A}$. Then $\delta(p, x, c) = (q, y)$, and, by (1), the final point of the walk (c, x) is $\ge y$, and so is in $\mathbf{N}^m$. The same is true for all the left factors c' of c. Hence, c is admissible for x, and so it is in L.

Conversely, let c be in L: we argue by induction on the length of c. We write $c = c'a$, with $a \in A$. Then $c' \in L$ and, by the induction hypothesis, $\delta(p,x,c') \neq \emptyset$. Set $\delta(p,x,c') = (q,y)$. It is enough to show that $\delta(q,y,a) \neq \emptyset$. Let i be a coordinate. If there exists no left factor c_1 of c' such that $x_i + v(c_1)_i > N$, then, by construction of $\mathcal{A}$, we have $y_i = x_i + v(c')_i$. But c is admissible for x, so a is admissible for the final point $x + v(c')$ of the walk (c',x); thus, $x_i + v(c')_i + v(a)_i \geq 0$, i.e., $y_i + v(a)_i \geq 0$. If, on the other hand, there exists a left factor, c_1 of c' such that $x_i + v(c_1)_i > N$, then by (2) there also exists a left factor c_1 of c' such that $\emptyset \neq \delta(p,x,c_1) = (r,z_i)$ with $z_i = N$. We choose a left factor c_1 of c' with maximum length such that $\delta(p,x,c_1) = (r,z_i)$ and $z_i = N$. We then have $c' = c_1 c'_2$ and $c = c_1 c_2$ (where $c_2 = c'_2 a$): lemma 6.2 implies that $v(c_2)_i \geq -N$. By (3), we then have $y_i = z_i + v(c'_2)_i$, from which we obtain $y_i + v(a)_i = z_i + v(c_2) \geq N - N \geq 0$.

In conclusion, $y + v(a) \geq 0$, which implies that $\delta(q,y,a) \neq \emptyset$, since $\alpha(a) = \omega(c') = q$ by (1), so $\delta(p,x,c) \neq \emptyset$, and c is recognized by $\mathcal{A}$. □

The language associated with a marked Petri net (R, M_0) is defined in a way analogous to that for a VASS: letting T be the set of transitions of R, this language is the set of transition sequences considered as words over T^* which are fireable for M_0. We have a simple correspondence between Petri net languages and VASS languages (cf. exercise 6.1): the recognizability problem for these two language classes are therefore equivalent.

6.2 An Undecidable Problem

We will say that a set K contained in $\mathbf{N}^k$ is *realizable* if there exists a VASS $G = (Q,A)$ labeled by $v: A \to \mathbf{N}^m$ with $k \leq m$, an initial configuraion (p,x) and a state $q \in Q$ such that:

$$K = \{y \in \mathbf{N}^k \mid \exists z \in \mathbf{N}^m, (p,x) \Rightarrow (q,z) \text{ and } y = \pi(z)\}$$

where π is the projection $\mathbf{N}^m \to \mathbf{N}^k$, $(a_1,\ldots,a_m) \mapsto (a_1,\ldots,a_k)$. We then have the following theorem:

Theorem 6.3. *The inclusion of realizable sets is undecidable.*

To prove this theorem, we reduce it to Hilbert's tenth problem which is known to be undecidable: given a polynomial $P(t_1, \ldots, t_k)$ with coefficients in $\mathbf{Z}$, the problem of finding a k-tuple $(a_1, \ldots, a_k) \in \mathbf{N}^k$ such that $P(a_1, \ldots, a_k) = 0$ is undecidable (theorem due to Matijacevič et al., cf. Manin (1977), theorem VI.1.2).

The basic idea is to simulate the computation of $P(a_1, \ldots, a_k)$ using a VASS. We know that any recursive function can be computed by a counter automaton. It suffices to imitate this result, making use of the striking analogy between counter automata and VASSs: the latter are counter automata from which the possibility of testing whether a counter is empty has been removed.

We will say that a function $f\colon \mathbf{N}^k \to \mathbf{N}^l$ is *weakly computable* by a VASS G (labeled in $\mathbf{Z}^m$) if there exist in G the states p and q (called the *initial* and *final* states), and sets of coordinates $I = \{i_1, \ldots, i_k\}$ and $J = \{j_1, \ldots, j_l\}$ (called the *entry* and *exit* coordinates), such that the following two conditions are satisfied:

(i) For all $x \in \mathbf{N}^m$, $y \in \mathbf{N}^m$ such that $i \notin I \Rightarrow x_i = 0$, and such that $(p, x) \Rightarrow (q, y)$, we have $(y_{j_1}, \ldots, y_{j_l}) \leq f(x_{i_1}, \ldots, x_{i_k})$ with the natural ordering over $\mathbf{N}^l$.

(ii) For all $(b_1, \ldots, b_l)$ in $\mathbf{N}^l$ and $(a_1, \ldots, a_k) \in \mathbf{N}^k$ satisfying $(b_1, \ldots, b_l) = f(a_1, \ldots, a_l)$, there exist $x, y \in \mathbf{N}^m$ such that $i \notin I \Rightarrow x_i = 0$, $t \in \{1, \ldots, k\} \Rightarrow x_{i_t} = a_t$ and $s \in \{1, \ldots, l\} \Rightarrow y_j = b_s$ (for j indexed by s), with $(p, x) \Rightarrow (q, y)$.

Example 3

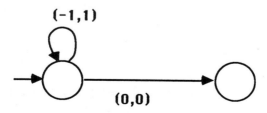

(-1,1)

(0,0)

Figure 6.4

The function $(x,y) \to x+y$ is weakly computable by the VASS shown in figure 6.4. For this VASS the entry coordinates are $\{1,2\}$ and the exit coordinate is $\{2\}$. The initial and final states are shown as inward and outward arcs. In fact, this function is also weakly computed by the VASS shown in figure 6.5.

■

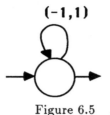

(-1,1)

Figure 6.5

A function $f:\mathbf{N}^k \to \mathbf{N}^l$ such that each coordinate $f(x)$ is given by a polynomial with natural coefficients, we call a *polynomial function*. We will show that such a function is always weakly computable by a VASS. To do this, we introduce particular functions and operations over them. It will be convenient to consider the set $\mathbf{N}^0 = \{0\}$.

    (i) *Sum:* $s:\mathbf{N}^2 \to \mathbf{N}$, $(x,y) \mapsto x+y$.

    (ii) *Product:* $p:\mathbf{N}^2 \to \mathbf{N}$, $(x,y) \mapsto xy$.

    (iii) *Constant function:* $\mathbf{N}^0 \to \mathbf{N}$, $c_0:x \mapsto 0$ and $c_1:x \mapsto 1$.

    (iv) *Deletion:* $\pi:\mathbf{N} \to \mathbf{N}^0$, $x \mapsto 0$.

    (v) *Transposition:* $t:(x,y) \mapsto (y,x)$; $\mathbf{N}^2 \to \mathbf{N}$.

    (vi) *Diagonal:* $\delta:\mathbf{N} \to \mathbf{N}^2$, $x \mapsto (x,x)$.

    (vii) *Identity:* $i:\mathbf{N} \to \mathbf{N}$, $x \mapsto x$.

    (viii) *Cartesian product* of two functions f and g: it is the function $g \times f:(x,y) \mapsto (g(x),f(x))$.

    (ix) *Composition:* gf (or $g \circ f$) of two functions g and f.

The classic result is left as an exercise for the reader (cf. exercise 6.6; see also Eilenberg and Elgot, 1970). Its proof will simply be illustrated by an example.

Lemma 6.4. *The class of polynomial functions coincides with the class of functions obtained on the basis of the seven functions (i)–(vii) by the application of a finite number of the operations (viii) and (ix).*

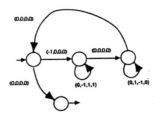

Figure 6.6

Example 4

The function $(x, y) \mapsto (xy, x+y)$, $\mathbf{N}^2 \to \mathbf{N}^2$ is equal to $(p \times s)(i \times t \times i)(\delta \times \delta)$. So, we have indeed $(p \times s)(i \times t \times i)(\delta \times \delta)(x, y) = (p \times s)(i \times t \times i)(x, x, y, y) = (p \times s)(x, y, x, y) = (xy, x + y)$. The function $(x, y) \mapsto 1 + x^2 y + y^2$ is equal to $s(i \times s)(p(i \times p) \times p \times c_1)(i \times i \times i \times \delta)(\delta \times \delta)$, as the reader may verify. ∎

Lemma 6.5. *Every polynomial function is weakly computable by a VASS.*

PROOF. Sum is weakly computable by example 3. Product is weakly computable by the VASS shown in figure 6.6. The entry coordinates are 1 and 2, and the exit coordinate is 4. To show that this VASS weakly computes the product, we observe that it is obtained from the following counter automaton (figure 6.7) which computes the function $(x, y) \mapsto xy$, where θ_i denotes the fact that the corresponding transition can only be used if the ith counter is empty.

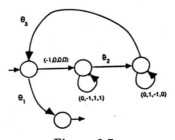

Figure 6.7

When one starts this automaton on the entry $(x, y, 0, 0)$, it terminates with the vector $(0, y, 0, xy)$.

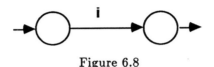

Figure 6.8

The constant function c_i is weakly computed by the VASS in figure 6.8 with an empty set of entry coordinates, and 1 as the exit coordinate.

Figure 6.9

Deletion is weakly computed by the VASS in figure 6.9, which has 1 as its entry coordinate, and $\emptyset$ as its set of exit coordinates.

Transposition is computed by the same VASS when 1 and 2 are taken as the entry coordinates, and 2 and 1 as the exit coordinates.

Figure 6.10

Diagonal is computed by the VASS in figure 6.10 with 1 as the entry coordinate, and 2 and 3 as the exit coordinates.

Figure 6.11

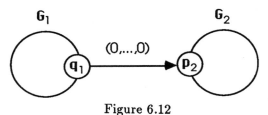

Figure 6.12

Identity is computed by the VASS in figure 6.11.

If f_1 and f_2 are weakly computed by the VASSs G_1 and G_2, labeled, respectively in $\mathbf{N}_1^m$ and $\mathbf{N}_2^m$, we begin by labeling both of them in $\mathbf{N}_1^m \times \mathbf{N}_2^m$, completing the vectors with zeros; then we connect them in series in the way shown in figure 6.12, where q_1 is the final state of G_1 and p_2 is the initial state of G_2. Then this VASS weakly computes $f_1 \times f_2$.

Now, let f_1 and f_2 be as above, and let them be composable. So, if J_1 is the set of exit coordinates of G_1 and I_2 the set of entry coordinates of G_2, we have $|J_1| = |I_2|$. To simplify, we can assume that $J_1 = J_2 = K$. Then G_1 is labeled in $\mathbf{N}^K \cup A$ and G_2 in $\mathbf{N}^K \cup B$. We then label each of these two VASSs in $\mathbf{N}^K \cup A \cup B$ by completing the labels by adding zeros. Next, we connect G_1 and G_2 as in the above figure. Then, if f_2 is increasing (i.e., $x \le y \Rightarrow f_2(x) \le f_2(y)$), the new VASS weakly computes $f_2 \circ f_1$.

The lemma is then proved using lemma 6.4 and by the fact that polynomial functions are increasing. □

PROOF OF THEOREM 6.3. 1. Let $P(t_1, \dots, t_k)$ be a polynomial with coefficients in $\mathbf{N}$. By lemma 6.5, the function $(a_1, \dots, a_k) \mapsto P(a_1, \dots, a_k)$ is weakly computable by a VASS G labeled in $\mathbf{N}^m$, with initial and final

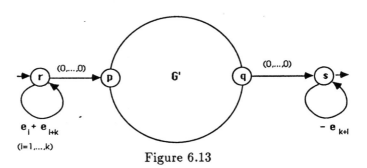

Figure 6.13

states p and q, and entry coordinates $\{1,\ldots,k\}$, and with exit coordinate l.

Then consider the VASS labeled in $\mathbf{N}^t \times \mathbf{N}^m$ as shown in figure 6.13, where e_i is the ith vector in the canonical base of $\mathbf{N}^{k+m} = I^k \times \mathbf{N}^m$, and where G' is obtained from G by replacing each label $v \in \mathbf{N}^m$ by $(0,\ldots,0,v) \in \mathbf{N}^{k+m}$. Let π be the projection from $\mathbf{N}^{k+m}$ onto the coordinates $1, 2, \ldots, k$ and $k + l$. Then the set (realized by G)

$$\{y \in \mathbf{N}^k + 1| \exists z \in \mathbf{N}^m, (r,(0,\ldots,0)) \Rightarrow (s,z) \text{ and } \pi(z) = y\}$$

is equal to

$$E(P) = \{(a_1,\ldots,a_k,a) \in \mathbf{N}^k + 1| \ a \leq P(a_1,\ldots,a_k)\}$$

2. Now, let P be a polynomial with coefficients in $\mathbf{Z}$. We can assume that $P(a_1,\ldots,a_k) \in \mathbf{N}$ for all $(a_1,\ldots,a_k) \in \mathbf{N}^k$ (we replace P by P^2). We can write $P = P_1 - P_2$, where P_1 and P_2 have coefficients in $\mathbf{N}$. Since $P(x) \geq 0$ for all $x \in \mathbf{N}^k$, we have $P_2(x) \leq P_1(x)$. It follows that we have $\forall x \in \mathbf{N}^k, P(x) \neq 0 \Leftrightarrow \forall x \in \mathbf{N}^k, P_2(x)+1 \leq P_1(x) \Leftrightarrow E(P_2+1) \subset E(P_1)$. This shows that if the inclusion problem for realizable sets were decidable, then so would Hilbert's tenth problem. □

Corollary 6.6.   *The inclusion of accessibility sets for Petri nets is an undecidable problem.*

We need a preliminary result. We will say that a set $K \subset \mathbf{N}^k$ is *realizable* by a Petri net if there exists a net $R = (P, T, Pre, Post)$, an initial marking M_0, and a subset $\{p_1, \ldots, p_k\}$ of P such that K is equal to the set of k-tuples $(M(p_1), \ldots, M(p_k))$ for all markings M accessible from M_0.

Lemma 6.7. *A set is realizable by a Petri net if and only if it is realizable by a VASS.*

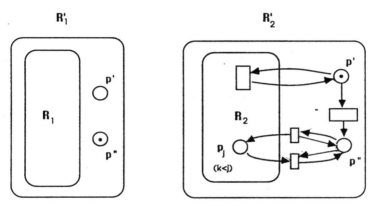

Figure 6.14

PROOF. It follows immediately from the construction in section 2.1, where it is shown that a Petri net can simulate a VASS and vice versa.
□

PROOF OF COROLLARY 6.6. For a marked net (R, M_0), we write its accessibility set as $A(R, M_0)$. We show that the inclusion problem for realizable sets in Petri nets reduces to the inclusion problem for accessibility sets. Let K_1 and K_2 be two subsets of $\mathbf{N}^k$ which are realizable by the Petri nets (R_1, M_1) and (R_2, M_2). We can assume that R_1 and R_2 have the same place set $P = \{p_1, \ldots, p_m\}$ (by adding to one of the net places which are connected to no transition). We can also assume that K_1 and K_2 are obtained by projections from $A(R_1, M_1)$ and $A(R_2, M_2)$ onto the places $p_1, \ldots, p_k$. We will construct two new marked nets (R'_1, M'_1) and (R'_2, M'_2) with common place set $P \cup \{p', p''\}$ as shown in figure 6.14.

The markings M_1' and M_2' coincide, respectively, with M_1 and M_2 on the places in P; they are shown in figure 6.14 for $\{p', p''\}$. A transition $\neq \theta$ in R_2' is only fireable if p' is not empty. Furthermore, if p'' is not empty, the places in $P \backslash \{p_1, \ldots, p_k\}$ can be filled with an arbitrary number of tokens. This shows that:

$$A(R_2', M_2') = K_2 \times \mathbf{N} \times \cdots \times \mathbf{N} \times \{0\} \times \{1\} \cup L$$

where we identify the marking and the vector it defines in n $\mathbf{N}^P \cup \{p', p''\}$, and where L is a set of markings such that $M(p') = 1$, $M(p'') = 0$.

More simply, we also have that the projection of $A(R_1', M_1')$ onto the places $p_1, \ldots, p_k$ is equal to K_1, and to $(0, 1)$ at places p', p''. Thus we have the following:

$$K_1 \subset K_2 \Leftrightarrow A(R_1', M_1') \subset A(R_2', M_2')$$

□

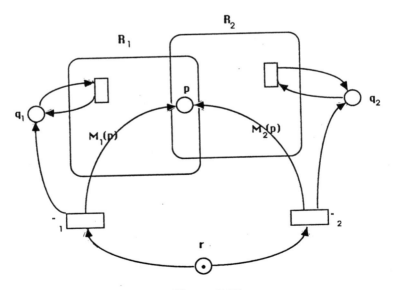

Figure 6.15

Corollary 6.8. *The equality of accessibility sets for Petri nets is an undecidable problem.*

PROOF. The inclusion problem reduces to that of equality. Indeed, let (R_1, M_1) and (R_2, M_2) be two marked nets having the same place set P. We construct three marked nets R, R_1' and R_2', having as place set, $P' = P \cup \{q_1, q_2, r\}$, and, as common initial marking, $M_0 = (0, \ldots, 0, 1)$, where we identify a marking with the vector it defines in $\mathbf{N}^P$. The net R is shown in figure 6.15.

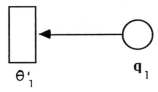

θ'_1 q_1

Figure 6.16

The transition θ_1 initializes R_1 and allows it to operate. The transition q_2 does the same thing for R_2. So, we have:

$$A(R, M_0) = \{M_0\} \cup A(R_1, M_1) \times (1, 0, 0) \cup A(R_2, M_2) \times (0, 1, 0)$$

The net R_1' is obtained from R by adding the transition θ_1' to it: this transition allows place q_1 to be emptied (figure 6.16).

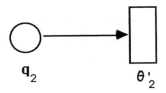

q_2 θ'_2

Figure 6.17

We thus have:

$$A(R'_1, M_0) = A(R, M_0) \cup A(R_1, M_1) \times (0,0,0)$$

Now, R'_2 is obtained from R'_1 by adding q'_2 as an extra transition θ'_2 which allows place q_2 to be emptied (figure 6.17).

We thus have:

$$A(R'_2, M_0) = A(R, M_0) \cup (A(R_1, M_1) \cup A(R_2, M_2)) \times (0,0,0)$$

As every marking in $A(R, M_0)$ ends in one of the triples $(0,0,1)$, $(0,1,0)$ or $(1,0,0)$, we have:

$$A(R'_1, M_0) = A(R'_2, M_0)$$

$$\Leftrightarrow A(R_1, M_1) = A(R_1, M_1) \cup A(R_2, M_2)$$

$$\Leftrightarrow A(R_2, M_2) \subset A(R_1, M_1)$$

□

6.3 Liveness

In a VASS with initial configuration (p, x), a transition a is said to be *live* if, for each configuration (q, y) which is positively accessible from (p, x), there exists a path including the edge a which starts at q and is admissible for y. We have already introduced the concept of a live transition in a Petri net in section 1.1. Thanks to the constructions in section 2.1, we can easily see that the liveness problem for Petri nets is equivalent to the liveness problem for a VASS. The object of this section is to show its decidability. To do this, we will say that a transition is *quasi-live* for the configuration (p, x) if there exists a path with start p which is admissible for x, and which contains the edge a. We can then say that a transition is live if and only if it is quasi-live for each configuration accessible from the initial configuration.

Lemma 6.9.   *Let $G = (Q, A)$ be a VASS labeled in $\mathbf{Z}^m$, p a state, and a an arc. The set of vectors x in $\mathbf{N}^m$ such that a is quasi-live for the configuration (p, x) is an ideal of $\mathbf{N}^m$, whose minimal elements can be effectively determined.*

PROOF. Let I be this set. If x is in I, and $y \geq x$, then y is also in I. Hence, I is an ideal of $\mathbf{N}^m$. For an element x of $\mathbf{N}^m$, the transition a is quasi-live for (p,x) if and only if the transition a appears in the Karp and Miller tree with initial configuration (p,x). This is derived from theorem 4.2 and lemma 4.4.

Let $\overline{I}$ be the ideal of $\mathcal{N}^m$ generated by I (cf. section 3.5). For $x \in \mathcal{N}^m$, we have $x \in \overline{I}$ if and only if there is a vector $y \in I$ which coincides with x on the set J of the finite coordinates in x.

But this is the same as saying that the transition a is quasi-live for (p,x_J) in the VASS obtained from G by projecting the labels onto $\mathbf{Z}^J$: this is decidable by the above. Hence, $\overline{I}$ is a recursive ideal of $\mathbf{N}^m$, and we can compute $\min(I)$ by theorem 3.12(ii). □

To finish we need one more theorem.

Theorem 6.10. *Let G_1 and G_2 be two VASSs labeled in $\mathbf{Z}^m$ with initial configurations (p_1, x_1) and (p_2, x_2) respectively. Let q_1 and q_2 be two states in G_1 and G_2. We can decide if the two accessibility sets $\{x \in \mathbf{N}^m | (p_1, x_1) \Rightarrow (q_1, x)\}$ and $\{x \in \mathbf{N}^m | (p_2, x_2) \Rightarrow (q_2, x)\}$ have a nonempty intersection.*

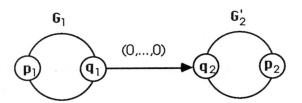

Figure 6.18

PROOF. It is enough to connect G_2 backwards onto G_1: let G_2' be the VASS dual to G_2 obtained by reversing the arcs and by replacing each label by its opposite. We then construct a new VASS as is shown in figure 6.18. Then the two sets of the theorem statement have a nonempty

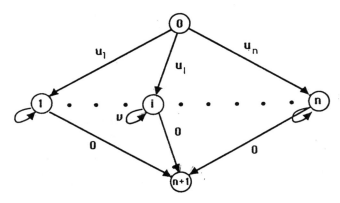

Figure 6.19

intersection if and only if, in G, the initial configuration (p_2, x_2) is accessible from (p_1, x_1). Therefore, this reduces to the accessibility problem which is decidable (cf. section 5.6). □

Corollary 6.11.   *One can decide if, in a VASS with an initial configuration, a transition is live.*

The same is thus true for Petri nets, as we saw at the start of this section.

PROOF. Let (p, x) be the initial configuration. The transition a is live if and only if, for each state q, and for each vector y such that $(p, x) \Rightarrow (q, y)$, a is quasi-live for (p, x). Set $I_q = \{y \in \mathbf{N}^m \mid a$ is quasi-live for $(q, y)\}$. The set I_q is an ideal for which we can effectively compute the minimal elements, and it becomes a matter of testing whether the set $\{y \mid (p, x) \Rightarrow (q, y)\}$ meets the complement $^c I_q$ of I_q. The complement $^c I_q$ is also an effective semi-linear set (ex. 3.18 of chapter 3). By theorem 6.10, it is enough to show, then, that each semi-linear set is a VASS accessibility set. This is immediate by an elementary construction. Let:

$$L = \bigcup_{i=1}^{n} (u_i + V_i^*)$$

We construct a VASS with $n+2$ states as shown in figure 6.19. Around each state $(1 \le i \le n)$, we put $|V_i|$ arcs labeled by vectors in V_i. □

Exercises

6.1 Let R be a marked Petri net with transition set T and an associated language L. Let T_1 and T_2 be two distinct copies of T, and let ϕ be the free monoid homomorphism $T^* \to (T_1 \cup T_2)^*$, defined by $\phi(t) = t_1 t_2$ (where $t_i \in T_i$ corresponds to the copy of T).

Show that L is recognizable if and only if $L' = \phi(L)$ is. Using one of the constructions in section 2.1, show that there exists a VASS whose associated language is $L'' = L' \cup \{wt_1 \mid wt_1 t_2 \in L'\}$. Show that L'' is recognizable if and only if L is.

6.2 Show that the VASS in exercises 4.3 and 4.4 do not satisfy the (VVN) condition.

6.3 Show that if a VASS with initial configuration (p, x) satisfies the (VVN) condition, then, for all states q, the set $\{y \mid (p, x) \Rightarrow (q, y)\}$ is semi-linear.

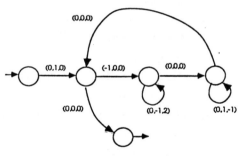

Figure 6.20

6.4 Show that the VASS in figure 6.20 weakly computes the function $n \mapsto 2^n$ with entry and exit coordinates 1 and 2, respectively.

6.5 Show that the VASS in figure 6.21 weakly computes the function $n \mapsto n^2$ with entry and exit coordinates 1 and 4, respectively.

6.6 Prove lemma 6.4. One can begin by showing that, for all k and p, and for all functions $\phi: \{1, \ldots, p\} \to \{1, \ldots, k\}$, the function $\mathbf{N}^k \to \mathbf{N}^p$, $(x_1, \ldots, x_k) \mapsto (x_{\phi(i)})$, $1 \le i \le p$, is in the class specified by the lemma.

6.7 Show that if R is a marked Petri net, and t is a transition, there exists a VASS G with an initial configuration and a transition a such that t is live for R if and only if a is live for G.

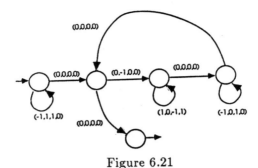

Figure 6.21

Notes

The results in section one are due, in their Petri net formulations, to Valk and Vidal-Naquet (1981): their extension to VASSs, as used here, is immediate. It is interesting to note that in order that a language for a VASS or Petri net be recognizable, it is enough for it to satisfy the iteration lemma for recognizable languages; or, alternatively, that its syntactic monoid be periodic (or a torsion monoid) – this follows, essentially, from the proof of theorem 6.1. Restivo and Reutenauer (1985) may be consulted for a study of this kind of problem. For a deeper study of the languages associated with Petri nets and their variants, see Peterson (1976), Crespi-Reghizzi and Mandrioli (1977), Mandrioli (1977), Starke (1978), Jantzen (1979), and Peterson (1981) and Schwer (1986, 1987). The results in sections 6.2 and 6.3 are due, in their Petri net formulations, to Hack (1974, 1976, 1979). We have chosen to present them in their VASS form because it is more readily manipulated. For the concept of a function which is weakly computed by a VASS, it is convenient to exploit the analogy between a VASS and a counter automaton. We know that every recursive function is computable by such an automaton; it is enough, then, in certain cases (e.g., the case of polynomial functions) to construct such an automaton, and to replace the emptiness test on counters by a transition labeled by the zero vector, and we obtain a VASS that weakly computes the function. For more on this, see also Müller (1985a). We have, however, followed Hack's original presentation of corollaries 6.6 and 6.8: the constructions used in their proof well illustrate the capabilities of Petri nets. Corollary 6.6 was proved by Rabin – see Baker (1973). Inclusion of realizable sets is decidable when they are finite (Karp and Miller, 1969, cf. theorem 4.9), but the complexity is considerable (for non-primitive recursive functions, see Mayr and Meyer, 1981). For other problems concerning decidability and complexity, see Jones, Landweber and Lien (1977), Rackoff (1978), Araki and Kasami (1977, 1977a).

References

Araki, T. and Kasami, T. (1977) Some decision problems related to the reachability problem for Petri nets, *Theoretical Computer Science*, **3** 85-104.

Araki, T. and Kasami, T (1977a) Decidable problems on the strong connectivity of Petri net reachability sets, *Theoretical Computer Science*, **4** 99-119.

Araki, T., Kagimasa, T. and Tokura, N. (1981) Relations of flow languages to Petri net languages, *Theoretical Computer Science*, **15** 51-75.

Baker, H. (1973) Rabin's proof of the undecidability of the reachability set inclusion problem for vector addition systems, *Computing Structures* memo **79**, Project MAC, MIT, Cambridge, MA.

Cardoza, E., Lipton, R. and Meyer, A. (1976) Exponential space complete problems for Petri nets and commutative semigroups, *Proceedings of the 8th ACM Symposium on the Theory of Computing*, 50-54.

Crespi-Reghizzi, S. and Mandrioli, D. (1977) Petri nets and Szilard languages, *Information and Control*, **33** 177-192

Eilenberg, S. (1974) *Automata, Languages and Machines*, vol. A, Academic Press, New York.

Eilenberg, S. and Schützenberger, M.P. (1969) Rational sets in commutative monoids, *Journal of Algebra*, **13** 173-191.

Eilenberg, S. and Elgot, C. (1970) *Recursiveness*, Academic Press, New York.

Fischer, M.J. and Rabin, M.O. (1974) Super-exponential complexity of Presburger arithmetic, in *Complexity of Computation, Proceedings of the SIAM-AMS Symposium on Applied Mathematics* (ed. Karp, R.M.)

von zur Gathen, J. and Sieveking, M. (1978) A bound on solutions of linear integer equalities and inequalities, *Proceedings of the American Mathematical Society*, **72** 155-158.

Ginsburg, S. (1966) *Mathematical Theory of Context-Free Languages*, McGraw-Hill, New York.

Ginsburg, S. and Spanier, E.H. (1964) Bounded Algol-like languages, *Transactions of the American Mathematical Society*, **113** 333-368.

Ginsburg, S. and Spanier, E.H. (1966) Semigroups, Presburger formulas and languages, *Pacific Journal of Maths*, **16** 285-296.

Ginzburg, A. and Yoeli, M. (1980) Vector addition systems and regular languages, *Journal of Computing and System Science*, **20** 277-284.

Grabowski, J. (1979) The unsolvability of some Petri net language problems, *Information Processing Letters*, **9** 60-63.

Grabowski, J. (1980) The decidability of persistence for vector addition systems, *Information Processing Letters*, **11** 20-23.

Hack, M. (1974) The recursive equivalence of the reachability problem and the liveness problem for Petri nets and vector addition systems, *Proceedings of the 15th Annual Symposium on Switching and Automata Theory*, IEEE, New York.

Hack, M. (1976) The equality problem for vector addition systems is undecidable, *Theoretical Computer Science*, **2** 77-95.

Hack, M. (1979) *Decidability Questions for Petri Nets*, Garland Publishing Co., New York.

Hopcroft, J. and Pansiot, J.-J. (1979) On the reachability problem for 5-dimensional vector addition systems, *Theoretical Computer Science*, **8** 135-159.

Huet, G. (1978) An algorithm to generate the basis of solutions to homogeneous linear diophantine equations, *Information Processing Letters*, **7** 144-147.

Huynh, T.D. (1982) The complexity of semilinear sets, *Elektr. Inform. Kybern.*, **18** 291-338.

Huynh, T.D. (1985) The complexity of the equivalence problem for commutative semigroups and symmetric vector addition systems, *Proceedings of the 17th STOC*, 405-412.

112   *References*

Huynh, T.D. (1985a) Complexity of the word problem for commutative semigroups of fixed dimensions, *Acta Informatica*, **22** 412-432.

Jantzen, M. (1979) On the hierarchy of Petri net languages, *RAIRO Informatique Théorique*, **19** 19-30.

Jantzen, M. and Valk, R. (1980) in *Net Theory and Applications* (ed. Brauer, W.), **84** 165-212, Springer-Verlag, Heidelberg.

Jantzen, M. and Bramhoff, H. (1983) in *Application and Theory of Petri Nets* (eds. Pagnoni, A. and Rozenberg, G.), Springer-Verlag, Heidelberg.

Jones, N., Landweber, L. and Lien, Y.E. (1977) Complexity of some problems in Petri nets, *Theoretical Computer Science*, **4** 277-299.

Karp, R.M. and Miller, R.E. (1969) Parallel program schemata, *Journal of Computer and System Science*, **3** 147-195.

Knuth, D.E. (1981) *The Art of Computer Programming, Volume 3, Seminumerical Algorithms*, Addison Wesley, Reading, MA.

König, D. (1950) *Theorie der endlichen und unendlichen Graphen*, Chelsea Publishing Co., New York.

Kosaraju, S.R. (1982) Decidability of reachability in vector addition systems, *Proceedings of the 14th Annual Symposium of the Theory of Computing*, 267-281.

Landweber, L.H. and Robertson, E.L. (1978) Properties of conflict-free and persistent Petri nets, *Journal of the Association for Computing Machinery*, **25** 352-364.

van Leeuwen, J. (1974) A partial solution to the reachability problem for vector addition systems, *Proceedings of the 6th Annual ACM. Symposium on the Theory of Computing*, 303-309.

Liu, L.Y. and Weiner (1970) A characterization of semi-linear sets, *Journal of Computing and System Science*, **4** 299-307.

Mandrioli, D. (1977) A note on Petri net languages, *Information and Control*, **34** 169-171.

Manin, Y.I. (1977) *A Course on Mathematical Logic*, Springer-Verlag, Heidelberg.

Mayr, E.W. (1981) An algorithm for the general Petri net reachability problem, *Proceedings of the 13th Annual Symposium on Theory of Computing*, 238-246.

Mayr, E.W. (1984) A algorithm for the general Petri net reachability problem, *SIAM Journal of Computing*, **13** 441-460.

Mayr, E.W. and Meyer, A.R. (1981) The complexity of the finite containment problem for Petri nets, *Journal of the Association for Computing Machinery*, **28** 561-576.

Mayr, E.W. and Meyer, A.R. (1982) The complexity of the word problem for commutative semigroups and polynomial ideals, *Advances in Mathematics*, **45** 305-329.

Müller, H. (1980) Decidability of reachability in persistent vector replacement systems, *Ninth Symposium on the Mathematical Foundations of Computer Science*, Lecture Notes in Computer Science, **88**, Springer-Verlag, Heidelberg, pp. 426-438.

Müller, H. (1981) On the reachability problem for persistent vector replacement systems, *Computing Supplements*, **3** 89-104.

Müller, H. (1982) Filling a gap in Kosaraju's proof for the decidability of the reachability problem in VAS, *Newsletter of the Special Interest Group 'Petri Nets and Related System Models'*, Gesellschaft für Informatik, Bonn.

Müller, H. (1985) in *Advances in Petri Nets 1984*, Lecture Notes in Computer Science, **188**, Springer-Verlag, Heidelberg, pp. 376-391.

Müller, H. (1985a) Weak Petri nets computers for Ackermann functions, *Elektr. Inform. Kyber.*, **21** 236-246.

Peterson, J.L. (1976) Computation sequence sets, *Journal of Computer and System Science*, **13** 1-24.

Peterson, J.L. (1981) *Petri Net Theory and the Modeling of Systems*, Prentice Hall, Englewood Cliffs, NJ.

Petri, C. (1962) Kommunikationen mit Automaten, Ph.D. Dissertation, University of Bonn.

Petri, C. (1962a) Fundamentals of a theory of asynchronous information flow, *Proceedings of the IFIP Congress*, North Holland, Amsterdam, pp. 386-390.

Petri, C. (1974) Introduction to general net theory of processes and systems, Hamburg; also in: Lecture Notes in Computer Science, **84**, Springer-Verlag, Heidelberg, 1980.

Rackoff, C. (1978) The covering and boundedness problems for vector addition systems, *Theoretical Computer Science*, **6** 223-231.

Reisig, W. (1982) *Petrinetze: eine Einführung*, Springer-Verlag, Heidelberg.

Reisig, W. (1985) *Petri Nets: An Introduction*, Springer-Verlag, Berlin.

Restivo, A. and Reutenauer, C. (1985) Rational languages and the Burnside problem, *Theoretical Computer Science*, **40** 13-30.

Sacerdote, G.S. and Tenney, R.L. (1977) The decidability of the reachability problem for vector addition systems, *Proceedings of the 9th Annual ACM Conference on the Theory of Computing*, 61-76.

Schwer, S. (1986) On the rationality of Petri net languages, *Information Processing Letters*, **22** 145-146.

Schwer, S. (1987) VASS Languages and Context-freeness, unpublished.

Starke, P. (1978) Free Petri net languages, *Proceedings of the 7th Symposium on the Mathematical Foundations of Computer Science*, Lecture Notes in Computer Science, **64**, Springer-Verlag, Heidelberg, pp. 506-515.

Starke, P. (1981) *Petrinetze*, Veb. Deutscher Verlag der Wissenschaften, Berlin (DDR).

Straubing, H. (1983) in *Combinatorics on Words: Progress and Perspectives* (ed. Cummings, L.J.), Academic Press, New York, pp. 279-295.

Valk, R. and Vidal-Naquet, G. (1981) Petri nets and regular languages, *Journal of Computer and System Science*, **23** 299-325.

Valk, R. and Jantzen, M. (1985) The residue of vector sets with applications to decidability problems for Petri nets, *Acta Informatica*, **21** 643-674.

Index